Industrial Property Markets in Western Europe

INDUSTRIAL PROPERTY MARKETS IN WESTERN EUROPE

Edited by Barry Wood and Richard Williams
Department of Town and Country Planning
University of Newcastle upon Tyne

Taylor & Francis
Taylor & Francis Group
LONDON AND NEW YORK

Published by Taylor & Francis
2 Park Square, Milton Park, Abingdon, Oxfordshire OX14 4RN
711 Third Avenue, New York, NY 10017

First edition 1992

Phototypeset in 10.5/12.5pt Garamond by Intype, London

First issued in paperback 2011

A catalogue record for this book is available from the British Library

Library of Congress Cataloging-in-Publication data
The Industrial property markets in Western Europe / edited by Barry Wood and Richard Williams.
p. cm.
Includes bibliographical references and index.

1. Real estate business—Europe. 2. Industrial buildings—Europe. 3. Commercial buildings—Europe. 4. Industrial site—Europe
I. Wood, Barry (Barry D). II. Williams R. H. (Richard Hamilton), 1945– .
HD588.I53 1992
333.33'8722'094—dc20 92–4881
CIP

Publisher's Note
The publisher has gone to great lengths to ensure the quality of this reprint but points out that some imperfections in the original may be apparent

ISBN 978-0-419-17050-1 (hbk)
ISBN 978-0-415-51189-6 (pbk)

CONTENTS

CONTRIBUTORS

Enrique Calderon is Acting Professor of Regional Planning in the Department of Regional, Urban and Environmental Planning, Escuela Tecnica Superior de Ingenioros de Caminos, Polytechnic University of Madrid.

Rocco Curto is Associate Professor in the Dipartimento Interateneo Territorio (Department of Territorial Planning, Faculty of Architecture), Politecnico di Torino.

Hartmut Dieterich is Professor of Vermessungwesen und Bodenordnung (Real Estate and Land Surveying), Fakultät von Raumplanung (Faculty of Spatial Planning), University of Dortmund.

Egbert Dransfeld is Research Associate in the Faculty of Spatial Planning, University of Dortmund.

Ignacio Espanol is Assistant Professor of Environmental Planning in the Department of Regional, Urban and Environmental Planning, Escuela Tecnica Superior de Ingenieros de Caminos, Polytechnic University of Madrid.

Alex Fubini is Associate Professor of Planning Theory in the Dipartimento Interateneo Territorio (Department of Territorial Planning, Faculty of Architecture), Politecnico di Torino.

Bert Kruijt is Professor of Construction Economics in the University of Amsterdam and Professor of Spatial Economics in the University of Nijmegen.

Alberto Landi is a Research Associate in the Dipartimento Interateneo Territorio (Department of Territorial Planning, Faculty of Architecture), Politecnico di Torino.

Hans Mattsson is Professor in the Institut for Fastighetsteknik (Department of Real Estate Planning), Royal Institute of Technology, Stockholm.

Alain Motte is Professor of Urbanisme et Aménagement (Urban and Spatial Planning) in the Institute d'Urbanisme de Grenoble, Université de Grenoble II.

Barrie Needham is Senior Lecturer in the Vakgroep Planologie (Department of Physical Planning), University of Nijmegen.

Richard Williams is a lecturer in Town and Country Planning in the University of Newcastle upon Tyne.

Barry Wood is a lecturer in Town and Country Planning in the University of Newcastle upon Tyne.

ACKNOWLEDGEMENTS

The editors and contributors acknowledge with gratitude the assistance of the following: for Chapter 5, Dany Boniface, Nichole Brochier, Florence Beloin, the Auguste Thouard Group, the Grenoble Isère Development Company and the Grenoble Chamber of Industry and Commerce; for Chapter 7, Dick Rouwenhorst, Jan-Willem de Geus and the Amsterdam office of Jones Lang Wootton; for Chapter 9, Agnette Linn and students on the MA European Property Development and Planning course; and for invaluable assistance with the preparation of the typescript, Colin Wymer for his ability to convert a wide variety of disk formats into the word-processing system understood by the editors, and finally Sharon McTeer for her skill, patience and efficiency in typing and word-processing the final text.

PREFACE

The Single European Market and the date 1992 have caught people's imagination and alerted many to the need to prepare themselves to take advantage of the commercial opportunities on offer. This is as true of the real estate, planning and property development professions as it is of any other sector. Many of the leading firms and professional associations are actively preparing themselves for a Europe-wide role.

There is a wide variety in the law and practice relating to real estate among the different member-states of the European Community. Anyone seeking to operate professionally on this wider stage has much to learn. This book is an attempt to assist in this learning process. It has been prepared alongside an initiative establishing a consortium, known as PRODEST, to promote graduate-level educational programmes in European real estate planning.

The Department of Town and Country Planning at the University of Newcastle upon Tyne, in which the editors both teach, is playing a leading role in this consortium, and has established a one-year MA/ Diploma in European Property Development and Planning, with the first students completing in September 1991. The PRODEST consortium is administered from the Politecnico di Torino, Italy, where another Masters course has been operating since 1989. Several of the other University departments represented by the contributors to this book are also members of the PRODEST consortium.

The idea for this book came from setting up these new courses, and from the realization that there was a severe lack of basic textbook material which took a European, as opposed to a national, perspective. There is a reasonable selection of urban and regional planning texts which take a European perspective, but very few on real estate. This book represents a first step towards remedying this situation.

Industry was chosen as the sector on which to concentrate first because, compared with the housing, retail and office sectors, the gap in existing material was greatest. Indeed, several of the people invited to contribute reported that this sector was largely devoid of existing coverage at their respective national levels and in their own languages. Thus, it is hoped that this book represents a useful addition to the textbook material available to students in each country represented.

The use of English as the medium for publishing a book such as this is not merely for the obvious reasons of being the editors' and publishers' language: it is also the only language held in common between all contributors and was acknowledged by all to be the most practical choice for a book such as this. As far as possible, however, 'Euro-neutral' terminology has been adopted rather than specifically British usage: hence, for example, the use of the term 'real estate' rather than 'the surveying profession'. Also, in order to minimize the confusion and loss of precision which can arise when non-English terms are translated into English, technical terminology is generally given in the original language as well.

Certain other conventions from European Community practice have been adopted to reduce the extent of British ethnocentricity which is almost inevitable in editing a book such as this. EC usage has been followed, for example, in the standard abbreviations for national currencies; and in the order of the chapters covering the different countries, which are in alphabetical order by their names in their own language. No editorial judgement is therefore implied or intended regarding the relative importance of their industrial property markets.

The selection of material and degree of detail to include or omit is always difficult to judge, and the editors acknowledge the inevitability of some frustration. We hope nevertheless that it is of some value and has succeeded in going some way towards meeting the purposes for which it is intended.

The editors would like to thank all those whose encouragement has helped to make this book possible, and to pay tribute to the interest and enthusiasm shown by all those who have contributed chapters. Without this, nothing would have been achieved. Responsibility for the final text, however remains with the editors.

BDW
RHW
Newcastle upon Tyne

1

INTRODUCTION

Barry Wood and Richard Williams

Note. Throughout milliard = US billion.

The prospect of the arrival of the Single European Market (SEM) at the end of 1992, and the need to think 'European' and to operate professionally throughout the SEM, have motivated the preparation of this book. It attempts to adopt the perspective of 'l'Europe mon pays', and to help readers take the first steps towards acquiring a level of knowledge of other countries in Europe equivalent to the knowledge that any professional in the field of real estate and planning would take for granted in their own country. Our objective is to provide real estate, planning and property professionals with an introductory guide, assisting them to start the process of learning how to operate in the property market in major countries within western Europe other than their own.

This chapter explains the scope of the book, outlines its rationale, and sets out the definitions, parameters and editorial principles which have been adopted. The second chapter explores at greater length the issues and assumptions on which these are based. Subsequent chapters are devoted to individual countries; each chapter is written by academic experts working in the country concerned. The final chapter draws the threads together, identifies common themes, comparisons and contrasts, and draws general conclusions.

All countries in Europe have some form of spatial or land use planning, allocating land uses and forming the basis for decisions on whether to authorize development. All the countries considered are advanced industrialized economies, and most have considerable pressure on their urban land available for development. All have substantial areas of land allocated and developed for industrial land and premises. Similarities end as soon as one moves beyond this very broad level of generalization. The principles of the various planning systems vary greatly, as is evident from studies of the various urban and regional planning systems which have taken a European perspective, such as Williams (1984) and Davies *et al.* (1989). The legal basis and formal procedures for the authorization of development similarly vary widely (Garner and Gravells, 1987).

When attention is focused on the property markets in different sectors, an even more bewildering complexity soon becomes apparent. Furthermore, although there is now a reasonably wide selection of material available on the different planning regimes in western Europe, there is less available on land and property markets. A valuable overview of all 12 EC member states has been prepared by Edward Erdman (1990), and Hallett (1988) offers a critical comparative analysis, from a cross-national perspective, of land policy and the residential property sector.

In spite of the complications and impediments represented by this complexity, which in EC single-market terms constitute substantial non-tariff barriers, the real estate industry (including many UK-based firms of chartered surveyors) has been among those responding energetically to the prospect of the SEM. Several larger UK-based firms have, for example, established a network of offices in the major cities of other member states, or formed partnerships with firms from other countries. Real estate and property firms from other countries are of course also interested in doing likewise.

What can a book such as this offer? In spite of the well-established Europeanization of the property industry and professions referred to above, there is a lack of textbook material, or indeed of any readily available reading matter, that provides introductory guidance to the system in other countries in a balanced way on which a deeper knowledge can be built with further reading. Much of the existing knowledge is held in the heads of the professionals operating in European firms, or is scattered in sources that are hard to identify from outside a country. New educational programmes are being developed to respond to this situation, including those referred to in the Preface to this book. Such courses are, however, quite new, and lack the textbook and literature infrastructure which support older-established teaching programmes.

With this book, we are taking the first step towards remedying this situation. Why Europe? This is clear from the above discussion. Why industry? This may be less so. The reason is that wealth creation through industry remains the lifeblood of the west European economy, yet there is less material on real estate development in this sector than there is on the retail, office and housing sectors. Consequently the gap to be filled is greater. The objective of this book, therefore, is to meet the need for the textbook identified above, to provide initial guidance to the respective national industrial property markets, to explain the terminology in the countries concerned, and their major planning, regulatory and financial controls, to explore the extent to which there is a true market, to describe the various actors (professional, political, commercial etc.) and operators in the industrial development process, and to provide a basis for further reading.

The focus is primarily on the processes of land use planning and authorization of development; on the terminology, concepts, rules and regulations of property and tenure; and on the operation of the market and the role of the actors or participants in the process. However, although detailed information on financial and tax issues would be of wide interest, such information becomes out of date so rapidly that it is inappropriate for a book such as this to attempt to provide it.

The scope of planning is understood to be, for the purposes of this book, the process of managing and regulating the use of land, space or territory by public authorities at the regional, urban or local scale in accordance with established policies or plans indicating the manner in which land should be used and urban and regional development carried out. Planning also includes the process of authorization of development and the use of land.

Planning is notorious for the extent to which its terminology is country-specific, reflecting the particular national planning culture to which it relates (Cropper, 1986; Williams, 1989). For present purposes, it is taken to include all aspects of spatial planning (*aménagement, Raumordnung, urbanisme* etc) relevant to the operations of the individual property markets, the allocation of areas of industrial land use, and the authorization of industrial development.

Industrial property is defined in functional terms, to include all property capable of being traded or transferred from one occupier or industrial process to another. In principle, therefore, it refers to general-purpose industrial sheds capable of accommodating a wide range of manufacturing (metal bashing, food processing, clothing for example) plus storage, warehousing and stockholding. Excluded from the scope of this definition are specialist industrial plant and structures such as oil refineries, chemical plants, and steel mills. Although plant and structures such as these may be bought and sold, there is no real general market in such property.

The type of general-purpose industrial premises on which this book is concentrating is not necessarily, of course, exclusively occupied by industrial users in the strict sense. The borderline between industrial and retail, or between industrial and commercial or service sector occupiers, is often difficult to draw. Where it *is* drawn, in land-use planning regulations for example, the definitions and categories, and indeed the true meaning of the vocabulary used, tend to differ between country and country. Therefore, an inclusionary approach has been adopted here. What is important is the type of building rather than the precise definition of the sector of the occupier, because the focus of attention is the market in this particular form of real property. The projects described in the various national contributions and the case studies presented as illustrations demonstrate the variety of interpretations of what constitutes industrial property of the sort considered here, and the variety of uses acceptable in such premises in the different countries discussed.

In selecting countries for inclusion, and in inviting contributions, the main aim was to include a representative sample of countries from the different parts of Western Europe, including all the major investment markets. In general, and in keeping with the SEM orientation referred to above, the selection has been from European Community countries. Sweden is the exception to this, but it is a major west European economy and an important member of the European Free Trade Association (EFTA). Furthermore, it is likely to become more closely associated with the EC during the 1990s, either as a result of negotiating to join, or as a result of an association between EFTA and the EC to extend the SEM in the form of a wider European Economic Area. A secondary consideration in the selection of countries, referred to in the Preface, is the link with the network of university teaching programmes in European real estate, property development and planning within the PRODEST consortium, and the need for textbook material for such programmes.

The issue of language is referred to frequently in the text. It is an issue which must always be confronted. The rationale for writing in English is explained in the Preface, and is clear enough. It is sometimes felt by members of the real estate profession and others, especially in Britain, that because textbook material such as this is published in English with the acquiescence of fellow professionals in other European countries, all communications can take place in English without risk of misunderstanding. The editors do not share this view. Many terms which are precisely understood in the professional context of one country are not capable of exact translation, and therefore may become ambiguous in a cross-national context. Terms in real estate and planning reflect the culture and practice of these activities in each country. If translated into English and back into the original language, they could be mistranslated, with loss of precision. For these reasons, international or Euro-neutral terminology is used where appropriate. The term 'Euroenglish' is sometimes used for the latter (Williams, 1989), to refer to the situation where English words are used to convey non-British concepts: thus, for example, 'spatial planning' instead of 'town and country planning'; 'real estate firms' instead of 'chartered surveyors'. Also, wherever desirable, the correct term in the respective national language is given in the text which follows.

REFERENCES AND FURTHER READING

Cropper, S. (1989) 'Do you know what I mean? Problems in the methodology of cross-cultural comparison', in Masser and Williams (Eds), *Learning from Other Countries: the cross-national dimension of urban and regional policy-making* (eds Masser and Williams), Geobooks, Norwich.

Davies, H. W. E. *et al.* (1989) *Planning Control in Western Europe*, HMSO, London.

Edward Erdman (1990) *Property*, Mercury Books, London, in association with the CBI Initiative 1992.

Garner, J. P. and Gravells, N. P. (eds) (1986) *Planning Law in Western Europe*, 2nd edn, Elsevier, Amsterdam.

Hallett, G. (1988) *Land and Housing Policy in Europe and USA*, Routledge, London.

Williams, R. H. (ed) (1984) *Planning in Europe: Urban and regional planning in the EEC*, George Allen & Unwin, London.

Williams, R. H. (1989) 'Are we speaking the same language? The vocabulary of planning in languages close to English', Paper to Conference of the Association of European Schools of Planning, Tours.

2

KEY ISSUES AND THEMES

Barry Wood and Richard Williams

2.1 GOVERNMENT AND PLANNING CONTEXT

Authorization of the use of land for development or economic activity normally needs to be sought from the appropriate public authorities. In all European countries industrial property development is only acceptable when subject to such authorization. In general, authorization is granted on the basis of conformity to some form of development plan or spatial planning policy. We need therefore to understand the concepts of authorization of development, of development plan and spatial policy formulation, and of the relationship between these two.

The normal model is an overall strategic policy framework, within which regional and local or municipal authorities prepare their own development plans and policies. Within areas designated for development, local and detailed site-specific plans are formulated and approved as a basis for authorizing development. In general, the legal basis for authorizing development is that if the proposal falls within the location and physical form specified in this detailed plan, it can, or must, be authorized. Very often, this detailed local plan is a mandatory element of the planning system, and authorization cannot proceed without such a plan having been legally adopted. Such plans are commonly referred to as binding plans. A legally binding form of plan is the normal basis of the planning system in most European countries, the main exceptions being the UK and Ireland.

The above model implies a strongly developed policy context at the regional and local scale, plus a deterministic relationship between the most detailed level of plan and the legal and administrative process of authorizing development. Indeed, the process of authorizing development in this model is sometimes seen to be purely an administrative process.

This very simplified basic model clearly does not apply equally to all countries being considered, but it will serve as a starting point from which certain key issues can be identified: issues which represent or indicate important differences in the ways in which different planning and authorization procedures operate, and their underlying assumptions and context. These issues include:

1. the spatial scale at which planning policy is formulated;
2. discretion on the part of the municipal planning authorities to decide whether or not to authorize development which does not fall exactly within the development indicated in the local site-specific development plan;
3. flexibility and certainty and whether the overall planning and development process is plan-led or market-led; and

4. the extent to which planning practice is characterized by negotiation prior to the formalization of any legally binding plan or other procedure for the authorization of development.

Flexibility on the part of the planning authorities to deviate from their previously existing planning policies in response to changing market conditions without going through highly structured legal procedures is often seen as the antithesis of certainty for the developer who expects to be able to know what form of development is acceptable and likely to receive authorization. The Dutch constitutional doctrine of *Rechtstaat*, whereby the citizen is entitled to legal certainty, is reflected here and in the basic model of the planning authorization process outlined above.

From these general issues, we move on to questions of the form of development plan, zoning, allocation of land uses, authorization and the control of development; and of the relationship between the planning system and the development process, including that of 'planning gain', 'developer obligations' or 'exaction' of a public good such as public infrastructure or amenity from the developer as part of an agreement by the public authority to authorize the development. For an extended discussion of this theme see Alterman (1988).

A final set of issues in the planning framework is the European Community context. European environmental legislation and the operation of the structural funds to promote development are important considerations to which attention is drawn by contributors wherever appropriate. A bigger underlying question is that of non-tariff barriers and impediments due to the variety of regulations, procedures and practices and professional roles in the real estate, planning and development processes in the different countries. The issue of harmonization of planning and development procedures, in order to facilitate the creation of the single market, has been raised in some countries.

2.2 SPATIAL SCALES OF PLANNING AND THE CONTROL OF DEVELOPMENT

The spatial scale at which planning policy is formulated varies considerably, as does the range of levels of government with planning responsibilities. Consequently, the locus of the political debate associated with spatial policy-making is found in a wide variety of circumstances, although it is invariably true that considerable political dispute surrounds planning decisions.

Whenever one is looking at spatial scales, however, it is always important to distinguish between a formal or legal requirement to prepare plans at a certain level in the national hierarchy of government (national, regional, provincial, conurbation authority etc.) and the existence of an actual plan or planning policy at that level. It is also necessary to distinguish between those plans which are mandatory in the sense that they must legally be prepared, or because the process of authorization of development cannot proceed legally without them, and those plans which are purely policy guidance, and where the appropriate authority may have legal discretion over whether to prepare such a plan.

To some extent there is a spatial policy context at the EC level. The structural funds, of which the most significant for industrial property is the European Regional Development Fund, have clearly defined areas of benefit, and the Regional Policy Directorate (DGXVI) is formulating a spatial policy framework for the Community, under the title 'Europe 2000' (Commission, 1991). Meanwhile, the Environment Directorate (DGXI) has issued the 'Green Book' on the urban environment (Commission, 1990) which seeks to promote sound environmental planning principles for urban development including, of course, industrial development throughout the EC.

Some countries have never really attempted spatial planning (as opposed to economic planning) on the national or regional scale. Others have planning laws requiring a full set of spatial plans at national, as well as regional or provincial level, but have not succeeded in developing an effective policy framework at this level and so have bypassed this legal provision.

Examples of effective spatial policy at the national level do exist, although regional policy frameworks are more widespread. The latter demonstrates a great variety in scale, scope, and political and legal backing. The explanation for this variety depends on another key variable in the countries being considered: the constitutional arrangements. These range from a federal system such as that of Germany with legislative power centred in the states, to systems of power devolved to regions either throughout the country or to certain historically distinct or separate parts (Italy and Spain). Others may have regional authorities with policy-making and executive powers, but no legislative power (France), and elsewhere regions may be merely administrative divisions of central government administration (UK) or non-executive associations of local authorities.

The power and significance of any regional planning policy depends on the political power and status of the regional authority, and also on the range of policy sectors over which it has jurisdiction. Another variable aspect of this is the spatial scale and number of regions into which a country is divided.

Similarly wide variations occur at the sub-region, city-region or conurbation scale of planning policy-making. The UK is the only country to have taken the radical step of abolishing existing conurbation planning authorities (the GLC and metropolitan counties) in 1986. Elsewhere such a level exists either as part of the regular hierarchy of sub-national government or as an *ad hoc* response to the need to have some form of conurbation-scale planning.

At the local scale, there are two types of plan which may typically be found: the land-use plan for the whole of a municipality or urban area, and a physical development or building plan at the scale of the precinct or even single development project. In several countries the latter is mandatory and an essential legal step in the process of authorization of development (e.g. *bestemmingsplan* in the Netherlands or *Bebauungsplan* in Germany). In these cases, which occur only in Roman law countries, the legal principle is that development proposals must be approved if they are in accordance with the plan, and cannot be authorized if they fall outside the framework of such a plan. This principle holds most strongly in those countries with a formally codified system of law, such as Germany and France. In these cases, the process of authorizing development is in principle an administrative process of checking conformity, not a process involving professional and political judgement and the exercise of discretion. In practice, of course, the rigidity implied by this description cannot and does not operate, and considerable flexibility is achieved by negotiation and other procedures.

The alternative model of local plans and their relationship to the authorization of development is to be found in the common-law countries of the UK and Republic of Ireland. In the UK the mandatory level of plan is at a quite different point in the spatial hierarchy, at the county (sub-regional) or city (whole municipal) levels. These plans are necessarily less precise, but are associated with a discretionary system for the control of development which in principle offers flexibility, but a loss of certainty for the developer.

One of the central attributes of such a system is that it offers both the public authorities and the developer scope for negotiation. The process can involve the effective sale of planning permission in exchange for certain public infrastructure or facility. It does however enable the planning system to respond to changing market pressures.

In the concluding chapter we shall be considering whether these two conceptually and legally quite different approaches to the authorization of development are in fact as far apart from each other in their practical effects for the industrial developer.

2.3 POLICY FORMULATION

Most introductory descriptions of national planning systems adopt the top-down structure used here. Implicit in this is the idea that policy is determined in a top-down manner from the establishment of broad parameters at the national or regional scale to the determination of acceptable development in a precise site-specific way at the most local level. This can of course be the case in some countries. In subsequent discussion we shall use the term 'policy-led' to describe such situations, where there truly is a system of policy formulation at these broader spatial scales, and where local plans and development decisions are taken in accordance within this policy framework.

However, a bottom-up concept may in reality apply more to the formulation of policy than this model implies. The broader spatial scales of plan may simply be the expression of the aggregate of local policies determined at the local level. It is important therefore to distinguish between the administrative appearance of a neat top-down policy-led planning framework and the reality, which may be much more *ad hoc* and responsive, representing the aggregate expression of local objectives. In addition to the top-down and bottom-up models of policy formulation, a third possibility exists, which we describe as 'market-led'. This refers to the situation where the planning system is designed to respond to the market in which developers operate, encouraging them to develop wherever they see the opportunity, with the plans offering guidance rather than direction, plus a procedure for authorizing development.

To return now to the context of the European single market, to which this book is addressed, a vital overall issue is the variety of planning and development authorization procedures. This variety constitutes a formidable non-tariff barrier for anyone in the real estate business. The question of whether planning procedures should be harmonized has therefore to be considered. Harmonization has been minimal in this field so far, although the Directive on the Environmental Assessment of Public and Private Projects of 1985 (Directive EEC/85/337), which has been in force from July 1988, is a significant step in this direction. This Directive also represents the first example of EC legislation substantively affecting the process of authorization of development (Williams, 1986; 1988). Any project for development which is likely to have significant environmental consequences must, before any decision is taken to authorize development, be subject to an environmental assessment. The range of projects which come under the terms of this directive are listed in two Annexes to the Directive. Several categories of industrial development come within its terms.

There is some interest in fuller harmonization of procedures for development authorization, although the upheaval involved would probably not be acceptable, unless it only applied to major categories of internationally significant or environmentally sensitive development. It is more likely that we will move towards development of the expertise to handle this variety. This is the perspective adopted by this book, and indeed its motivation.

2.4 PUBLIC POLICY CONTEXT

Having discussed the different models of planning policy framework to be found in the countries of Europe, we turn to the policies and political objectives conveyed by these different forms of plan and planning policy.

A spatial policy for industrial development may be formulated for a variety of purposes. The extent of government intervention in the market, and the directions in which developers are encouraged to look, depend not only on the political ideology of the government but also on the significance attached to different problems and the perception of the contribution that industrial development can make to overcoming them.

Broadly, political objectives are likely to fall into one of three categories:

1. policies designed to overcome disadvantage on the part of certain regions or urban areas;
2. policies to achieve a better quality environment; and
3. policies based on the objective of increasing national or regional wealth and competitiveness.

In any of these categories, policies for industrial development may have as their prime purpose that of making a political or ideological statement on the part of government. This was the case, for example, with Enterprise Zones in the UK. One important and traditional public policy objective is that of overcoming weakness in local economies and regional disparities. Thus industrial development is encouraged and possibly promoted by subsidy or subvention in economically weaker regions. These may be either regions that have never enjoyed advanced industrialization, or older industrial areas with outdated infrastructure and declining industrial sectors. The disparity category is sometimes extended to target cities, in an urban rather than regional policy, or certain types of urban situation, as in the UK inner city policy.

Specific objectives may be varied: job generation, or development of new sectors in the regional industrial economy which may be capital-intensive rather than labour-intensive, capture of internationally mobile investment using available financial incentives, or improvement of an area's negative image.

Environmental enhancement may be closely linked to image improvement, but it may also be a political objective in areas where the regional disparity argument does not apply, or is not adopted as a basis for policy. The adoption of environmental standards internationally at the EC level or on a wider basis is not only a necessary response to the problem of tackling environmental pollution: it is also closely related to economic development, harmonization and competition policy objectives within the EC. The argument is that in a single market, different national environmental standards applied through the various planning and authorization systems would amount to a distortion of competition, possibly leading to the existence of pollution havens into which certain categories of industrial development would be attracted by less costly environmental requirements. Harmonization of environmental standards, operating through the planning system, is therefore necessary to ensure a 'level surface' for the location decisions of industrial developers. This argument is developed in Williams (1986). The Environmental Assessment Directive of 1985 is an important first step in the direction of creating this level surface, but by itself is a long way from achieving this objective.

A further category, which is surely to become more important in the SEM, is improvement of national competitiveness by boosting the most competitive sectors of industry, by improving the national share of the newest and most advanced technologies (through science and technology parks and business parks etc.), and possibly also by promoting further development of existing industry in the most competitive sectors or locations. Such a policy, of course, is likely to have a different spatial expression from one addressing regional or urban disparities.

Another dimension of public policy in this context is the extent to which governments deem it proper to intervene in the land markets, assembly of land for industrial purposes, and provision of premises. For example, the concept of an advance factory, built by public authorities in order to encourage new industrial initiatives but in advance of any actual occupier being identifiable, has long been acceptable to governments in the UK and Ireland but only rarely elsewhere, and was for several years not acceptable as a project which could receive financial support from the ERDF. Other forms of intervention by municipal authorities, for example assembly of land, sites, and development of infrastructure for industry, have been much more readily accepted practice in some countries than others.

There are also variations in how far policies for industrial development to meet objectives such as those suggested above are seen as legitimately part of planning and therefore of spatial policy, and how far they are seen as part of separate, unrelated economic policy. The chapters which follow demonstrate this variety in public policy objectives in planning for industrial development.

2.5 CONCEPTS OF REAL PROPERTY

As stated earlier, this book is concerned with real industrial property, loosely defined as land and buildings used in manufacturing (and similar) processes. Real industrial property can be seen to fulfil two distinct roles in Europe. First it is part of the productive system, a means of production, as necessary as machinery and labour in the manufacture of industrial products. Indeed the existence of simple, usually single-storey buildings, operating as a sheltered location in which a wide range of activities takes place, can be identified in each European country. But the existence of buildings is only a necessary and not a sufficient condition for a market to exist. An occupier market also requires both a process of entry and exit from the market by occupiers and a price adjustment mechanism.

The process by which the right of occupation is obtained differs between countries. In some places, access will occur via ownership while elsewhere this is not essential, and in others outright ownership may not even be possible. Equally, price adjustment to changing patterns of demand can vary from 'free market' to 'state regulation'. A real European market in industrial property can only exist when potential entrants are aware both of the process of access (and exit) and the benefits and obligations associated with differing forms of occupation. One of the aims of this book will therefore be to define and clarify the context in which property occupation takes place in each of the countries reviewed.

Where the ownership of property can be separated from occupation there exists the potential for property to fulfil a second role, as an investment medium. This role is not essential: a property investment market where properties can be traded between owners without directly affecting occupiers is not an essential characteristic of an industrial society. Owner occupation of industrial property could be the normal practice. Where an investment market does exist the legal framework and the behaviour of market participants in each country will differ and the extent of the market and its characteristics will be determined by these differences. Another theme of this book will be the context in which property investment markets function.

The above comments have assumed that there are few problems in defining industrial property and, more importantly, concepts such as ownership and occupation. This assumption must now be dropped and replaced by a more rigorous approach to real property. This book will use a conceptual framework which will permit each national chapter to identify 'local' definitions and working practices. This market framework is British in origin and in the final chapter the editors will examine the extent to which this approach is sufficiently flexible to be adapted for European-wide usage.

It is conventional in Britain to argue that the property market is one where certain bundles of property rights are traded. These rights are:

1. to own property;
2. to occupy property;
3. to develop property;
4. to sell property;
5. to demolish property;
6. to change the use of property.

The rights are not self-evident: the meaning of 'ownership' or 'development' is intrinsically bound up with the laws and customs of each country and consequently one objective of this book must be to clarify the meaning of each of these rights within the context of each country's land laws.

Two rights, those of ownership and occupation, must be at the centre of any analysis of the operation of the industrial property market, because industrial property must be created by an organization and the investment of resources by that body must imply some form of ownership if returns are to be obtained. Equally the property would only be built if there were the expectation of occupation so that returns could be generated. A further right, to develop property, is also significant because industrial property deteriorates over time or becomes obsolete and will need to be replaced. This is also the major mechanism by which supply responds to changes in demand.

The right of ownership can have various meanings. In Britain the freehold ownership of land and buildings, where the rights exist indefinitely, is a well-defined concept. The extent to which the ownership includes mineral rights, extends above and below surface and includes alternative uses is defined in statute and subsequent case law. But other models exist. In some countries ownership of the land is held indefinitely by the state and the ownership of the property is restricted to a specified number of years.

Ownership may also be conditioned by the contractual relationship between the owner and an occupier, or between the site owner and adjacent site owners. Contracts differ between European countries and the interpretation of each is determined within each country's legal system. Furthermore, the ownership right is always conditioned by the state. This last point is crucial because questions such as 'does ownership include the right to demolish the existing premises and build a new property?' are highly significant in determining the value of a property and hence its investment potential. Ownership is nowhere absolute in the sense that it is unconditional, but equally the normally expected, or occasional but possible conditions differ greatly from country to country.

Similarly, occupation is not a self-evident concept. It may imply obligations associated with the repair and maintenance of the buildings, or with obligations to the owner, and the length and financial conditions of occupation may be regulated by statute. Additionally it may only be possible to occupy property by first obtaining the ownership rights. Where owner occupation is the norm the key issues are likely to be why a rental market has not developed, and the process by which access to property is obtained. If legislation restricts renting then not only do the legislation's detailed provisions need to be explained but also its objectives must be analysed. In this way the benefits and costs of the legislation can be analysed relative to other approaches. Legislation may not be the only cause of the non-existence of a rental market. Corporate behaviour differs between member states as do banking regulations and these are clearly relevant to the industrial property market. Even cultural norms of behaviour may be relevant, for the actors and agencies in each country have evolved their practices over time to meet the needs and conditions of their locality. Each national chapter therefore seeks to provide the background to enable this issue to be appreciated.

The definition of property development differs between EC countries. As a process it is highly dependent on the planning regime and also on the financial and business sectors in each country. Again legislation and normal patterns of behaviour interact to produce nationally distinctive systems which this book will aim to describe.

This book will therefore attempt to define and explain the property market concepts in each country. It will attempt to connect property and planning to give an overview of each country's property market system. The contributors will address the issues of occupation, ownership and development in each country in the form and structure in which they are normally understood locally. The reader should therefore be able to connect the key concepts with local practice and understanding.

One key characteristic of property, which complicates both the occupation and investment markets, is that property is non-homogeneous in a variety of ways. It is geographically fixed in location, is built to a number of quality standards, varies in age and physical depreciation, includes a variety of facilities and involves a range of management costs. When property is traded, further diversity results from the absence of centralized marketplaces, centralized information, and high transaction costs. At the European level, differing building regulations, local vernacular design and a host of 'normal' practices create even greater diversity. Hence, inevitably the industrial property market must be seen as disjointed and imperfect.

Local real property markets are not isolated. They do not operate in a vacuum, separated from other localities and countries, nor are they isolated from financial markets. As Europe progresses towards the SEM the interconnections between property markets will inevitably deepen, and certainly the establishment of a common currency would encourage property occupiers, developers, owners and market professionals to overcome geographical boundaries. In these circumstances imperfections become crucial to market participants and national governments.

Actors in the property development process need to recognize local practices if they are to make informed decisions on the location of industrial production and property investment. Governments will not wish to discourage inward investment or the creation and expansion of local business by operating a property market system which is unattractive to occupiers and/or investors.

Financial integration in Europe is proceeding rapidly. Banks and other financial intermediaries provide corporate finance at the European level to a growing number of companies which have integrated European operations. While decisions on the location of production will not be determined by property considerations alone, property nevertheless forms the bulk of fixed assets of many companies. At the level of direct property investment the financial institutions in Europe are increasingly able to diversify from home markets as restrictions on capital mobility are eased. The destination of these increasingly mobile funds will be influenced by the legislative framework and operating practices in each country.

2.6 PUBLIC SECTOR INTERVENTION

Planning is one way in which government intervenes in the independent functioning of the property market, but there are other forms of intervention. The state or one of its agencies may provide the land or buildings, may provide public subsidies to development or occupation activity, or may place financial restrictions on capital values or rent levels. Additional financial impacts can come from different tax regimes and levels. All these policies can collectively be described as financial intervention because they are targeted towards the costs and returns of investment and occupation. By comparison, while planning regimes and policies carry substantial financial implications, these are essentially a by-product of land-use policies formulated to achieve wider planning goals.

While financial intervention may be fiscally neutral, since government has no particular objective other than to bring real property into the same finance and tax regime as other corporate investment in the wider economy, it is apparent that the public sector throughout Europe controls, manipulates and supports the real industrial property sector. This far from neutral and clearly distorting behaviour is identified in this book where appropriate because it is difficult to comprehend local market behaviour without reference to public policy, although financial and taxation policies are beyond the scope of this book. The intervention can only be understood in terms of the goals of public policy to which reference was made earlier in this chapter and therefore another objective of this book is to indicate the purposes of each public policy instrument.

These issues, like several other themes in this book, are likely to be affected by the process of European integration, as the Commission will seek to harmonize subsidies in order to make the single market a reality.

2.7 THE FUNCTIONING OF INDUSTRIAL PROPERTY MARKETS

Markets exist to facilitate exchange, and the property market is no exception to this general rule. Occupiers and owners purchase property rights and in so doing relocate resources. The operation of any market can only be understood in relation to a set of norms concerned with the functioning of markets generally. Similarly optimal market behaviour can only be defined in terms of pre-determined criteria for markets generally. Any attempt by authors to evaluate their property markets needs to be related to these criteria.

In Britain the traditional model of the market is static: demand and supply can be identified and market price is assumed to adjust immediately to clear the market. This equilibrium model is perceived as a sort of ideal, for under competitive market conditions the market equilibrium can be shown to maximize consumer welfare. The market is said to fail if imperfections (or monopoly elements) distort behaviour. Inevitably much economic research has been directed towards measuring the costs of these imperfections, in terms of prices, output and overall economic welfare. This has resulted in much debate about the role of government in influencing market behaviour by restricting or regulating corporate activity and it has been used as a justification for state nationalization of private-sector monopolies.

A second approach to market behaviour is one which is dynamic. It perceives demand and supply decisions as unknown and unknowable: as a continuously changing flow of information. The market is at the intersection of changing patterns of consumer preferences and cost and availability information. The market is neither in nor out of equilibrium; rather it is a process by which conflicting pressures are efficiently resolved. This approach commonly leads to questions about the flow of information to the marketplace. Are flows efficient? Are they distorted by government policy and legislation? The approach accepts market imperfection as the norm and justifies the existence of markets in terms of the efficiency with which they respond to changing information flows. This book will attempt to evaluate the efficiency of each country's industrial property markets in terms of these norms.

The demands of property occupiers are central to the operations of the market in real industrial property. They each have particular property requirements which are met, with varying degrees of satisfaction, by the property which they currently occupy. New tenants, created either by relocation or by business formation, provide evidence on user requirements and indeed provide the stimulus for property development. The ways in which the requirements of new users are met, and the ways demand is stimulated in the different countries reviewed, are important indicators of the prevailing characteristics of the market in industrial property. The industrial property market must, above all else, be able to respond quickly to changes in demand if it is to be seen as efficient.

2.8 MARKET ADJUSTMENT

For the purposes of this book, market adjustment in the real industrial property market is defined as the process by which the existing stock of property is adjusted to meet new demands. This includes changes in property values and rents, changes in specification by refurbishment, addition and division, and changes in the use of properties. The distinction between market adjustment and market responsiveness is important as the supply of industrial property is dominated by the existing stock, and the ability of this to adjust to new demands will be limited.

2.9 MARKET RESPONSIVENESS

Market responsiveness is defined here as the adjustment of the industrial property market to meet new demands by the development of new buildings. This includes such issues as relocation (both within urban areas and between them), new building specifications and, where appropriate, the legitimization of changes in the designation of land for development.

The issue of development is inevitably complex because it sits at the crossroads of market demands and planning policies. The differences between national planning systems would be sufficient to ensure widely differing property development processes but differing forms of demand and their expression via differing economic and legal systems result in nationally distinctive property development markets.

It is then the functioning of industrial property markets upon which this book concentrates. There are, of course, variations within each country as well as major differences between each country reviewed. It is hoped nevertheless that the reader will be introduced to the essential elements of the process in a way that is both descriptive and evaluative, and will be equipped with the basic information on which to develop an operational knowledge of the various industrial property markets in western Europe.

This chapter has identified a range of activities and processes which may differ between European countries. In the final chapter the editors will attempt to identify some of the key themes, comparisons and contrasts which are identified in the national chapters.

REFERENCES AND FURTHER READING

Alterman, R. (ed) (1988) *Private Supply of Public Services: Evaluation of real estate exactions, linkage and alternative land policies*, New York University Press, New York.

Commission of the European Communities (1990) *Green paper on the Urban Environment*, Communication from the Commission to Council and Parliament, Brussels, COM (90) 218 final.

Commission of the European Communities (1991) *Europe 2000. Outlook for the development of the Community's territory: a preliminary overview*, Communication from the Commission to the Council and Parliament, Brussels and Luxembourg.

Williams, R. H. (1986) The EC environment policy, land use planning and pollution control. *Policy and Politics*, Vol **14** (1), 93–106.

Williams, R. H. (1988) The European Communities Directive on Environmental Impact Assessment, in *The Role of Environmental Impact Assessment in the Planning Process*, Ch. 4 (ed M. Clark and J. Herington) Mansell, London.

3

GERMANY

Hartmut Dieterich and Egbert Dransfeld

The operation of the industrial property market described here is basically that which has been developed in the western part of Germany, i.e. the Federal Republic before unification with the GDR. It is thus a system developed in the context of a highly industrialized and successful economy. However, the total population of Germany is now about 78 million, occupying 357 000 km. Reference is therefore made to the situation in the eastern part of the country (the former German Democratic Republic) in addition to describing the industrial property market as it has evolved in the western part.

There are big regional differences between the parts of Germany. In the western part the economy has been growing for eight years, and the rate of unemployment had fallen to 5.9% by September 1990. There is a substantial export surplus, and it is especially noticeable that the demand for industrial land and buildings is growing. Whereas the turnover of land decreased for all other uses in the years from 1980 to 1988, the turnover of industrial land grew within this period from 1200 ha to 2300 ha a year. New construction, as well as the improvement, of industrial estates is one of the major urban tasks, with an emphasis on good design and a desirable environment.

In the new *Länder* (states) in the eastern part of Germany the situation is rather dismal. The industrial output in these regions which are in transition from a Marxist command economy to the social market economy has decreased by 7% during the first half of 1990 and is still falling. Furthermore, construction activity is less than it was a year ago. Unemployment is rising, and it is still risky to buy property. It might belong to proprietors expropriated decades ago who could have their property restored to them. But it is to be expected that these difficulties will be overcome and that with the same legal framework similar tendencies of economic development will eventually be seen in all German *Länder*.

Industrial property is a term used in Germany in a very specific way. Speaking of small or middle-sized premises for manufacturing the term used is *Gewerbegrundstücke Industriegelände* (industrial property), because planning law differentiates between *Gewerbegebiete* (areas for commercial or trade property), which are to locate enterprises which do not seriously disturb their environment, and *Industriegebiete* (industrial areas). The latter are for the location of all kinds of industry, especially for factories not allowed elsewhere. Industry in Germany therefore means such activities as steel works, oil refineries, chemical plants, car factories and similar bigger estates. The expression *Gewerbe* (trade) is used not only for craft and workshops, but for all small industrial firms linked to manufacturing as a whole, operating in small premises; *Gewerbebetriebe* (commercial or trade businesses) can be found in industrial areas too. Industrial areas are of considerable quality today. Talk of trade and industry no longer conjures up images of dirty premises. We shall cover land and premises of better quality and of up to approximately 1000 m^2 floor space, whether they are industry, trade or commercial businesses according to the definitions used in Germany.

Section 3.1, dealing with the legal framework, will first outline the different kinds of occupation of industrial property according to civil law, and then discuss planning policy with regard to industrial property and examine the planning system. The influence of the planning system on the market for industrial land is enormous. Only if undeveloped land is designated by a *Bebauungsplan* (local plan; see below) for industrial use can it be traded as such. The influence of planning law on the property market is also important because changes in the use of property can be carried out only within the scope allowed by planning law. The price of land and property therefore depends to a great extent on the local plan.

The *Baugesetzbuch* (Planning Code or Town and Country Planning Act of 1987, abbreviated to BauGB) also contains rules for the first stages of implementing plans, especially by methods of land assembly; these will also be outlined, as well as the legal information system for land and property values. The section closes with a short survey of taxes and subsidies.

Section 3.2 discusses the market for industrial property. The land and development market, as the most important part of the industrial property market, will be dealt with, followed by the user market and the investment market.

Section 3.3 presents two case studies, and section 3.4 provides a variety of data.

3.1 LEGAL FRAMEWORK

Three areas of legislation are essential for the operation of the industrial property market:

1. civil law, providing the framework for occupation of industrial premises;
2. planning legislation;
3. taxes on land, buildings and businesses as well as subsidies for industry.

3.1.1 Occupation of industrial property

The legal framework for the occupation of industrial property can be divided into ownership, *Erbbaurecht* (hereditary leasehold), and *Miete* and *Pacht* (tenancy).

Owner occupation

Owner occupation is the most common form of tenure of industrial property. Most firms in Germany occupying industrial property will buy a piece of land and build the building with the help of their architects so that it can be designed specifically for their purposes. Sometimes a piece of property, i.e. land with building, is bought, enabling the owner to occupy the land and make use of the building.

Ownership is an exclusive right. The owner of a piece of real estate can do with it whatever he wants and can exclude everybody else from any influence on the property. The owner can use his property by himself; he can let it to tenants; he can sell it; he may even allow it to decay. He can also use it as a security for a loan. However, the right of ownership is not unlimited. The owner may do what he thinks fit only unless the law or rights of third parties, e.g. neighbours, do not stand against it. Very often planning legislation limits the owner's right to construct, to demolish, or to change the use.

Access to industrial land and property is free, and is only a question of price. If somebody buys land and constructs premises, or if somebody buys property with buildings, they can keep it indefinitely. There is no charge other than the taxes and fees due on all properties. The owner can change the use of the property, within the limits of planning law, and may erect another building if the old one is no longer of use.

Increases in land value arising during the time of ownership belong to the owner; they are not subject to taxation. Ownership is the kind of tenure most attractive for industrialists. In many cases ownership is unavoidable: the entrepreneur needs his property as security for the mortgage which is often necessary to start or to enlarge his business. Under these circumstances a major part of the market in industrial property is a market in land.

Erbbaurecht (hereditary leasehold)

A substitute for ownership is an *Erbbaurecht* (hereditary leasehold). An *Erbbaurecht* is acquired if someone cannot afford to pay the price for the land, or if the owner is not willing to sell his property permanently. The *Erbbaurecht* is the hereditary right to use the site and to have buildings on it. Buildings erected on a site with an *Erbbaurecht* are a component part of the *Erbbaurecht*, not of the land, and are owned by the person who bought the *Erbbaurecht*.

For industrial properties the duration of the right is usually 40–66 years. During these years the person who bought the *Erbbaurecht* enjoys almost all the rights of the owner. However, he has to pay a yearly rent to the owner. It is usual to arrange the average return on land of this kind as *Erbbauzins* (ground rent), typically about 6% or more of the value of the industrial land. In northern Germany it is also common to buy an *Erbbaurecht* with a single payment.

The holder of the *Erbbaurecht* can sell it. The value depends on the unexpired time of the right (and of course on the value of the buildings). The holder of the right has to pay taxes like an owner.

Only municipalities, large estates and the churches are establishing and selling *Erbbaurechte*; private persons and small investors seldom do so. *Erbbaurechte* are only a small sector of the industrial property market.

Miete/Pacht (tenancy)

Less popular in Germany is the occupation of industrial property as a tenant, i.e. via *Miete* or *Pacht*. However, tenancy of industrial property is becoming increasingly important: its significance has grown considerably within the last two decades.

A *Mietvertrag* or *Pachtvertrag* (lease) is a contract between owner and occupier. The rules for the lease are to be found in the *Bürgerliches Gesetzbuch* (BGB, 1896 – Civil Code, § 535 ff.) If the contract is for a longer period than one year it must be in writing. Such a contract may run for a finite number of years, but not for more than 30 years, with the exception that the contract shall run for the lifetime of the tenant or of the landlord. It is common to agree on a certain period, e.g. 10 years, with a clause that the contract will continue unless one of the parties would like to discontinue the lease, giving notice to quit some time in advance. If a lease is running for an indefinite time, each party may give three months' notice to quit at the end of a quarter of a year (unless a longer notice period has been agreed).

The tenant has to pay rent, usually monthly. Rents can be freely arranged in the absence of rent control for industrial property. Often the tenant is also required to pay for part of the maintenance and some additional costs, e.g. public fees for the property such as street cleaning and other public services. It is possible to agree a rent increase every few years in advance. No landlord is willing to bind himself to a given rent for too long. If there is a clause in the contract allowing for an increase of the rent according to the price of some commodity, to the index of industrial rents, or to inflation as a whole, then the permission of the Federal Bank is necessary.

If a building is occupied by more than one tenant, it is usual to arrange for service charges for common spaces at an additional cost. Such charges may be high in buildings, such as those in a business park, which enjoy many common facilities.

If the tenant does not fulfil his duties – to pay rent and to use the property correctly as agreed – the lease may be terminated without prior notice to quit.

The owner can sell the rented property to a third party. A sale does not affect the lease. The new owner has to meet the terms of the lease. This is also an advantage for the landlord. The tenant has to fulfil the contract, and a landlord is not hindered from capitalizing. A favourable lease may increase the value of a piece of property and be the reason why a good price is realized.

The tenant is not entitled to sublet the rented property without the consent of the owner. But if the owner does not consent to sublet, the tenant may terminate the contract with three months' notice to the end of any quarter of the year.

The rules for leases, whether *Miete* or *Pacht*, are quite clear and not difficult to handle. For investors, leasing offers the opportunity of a secure investment with a stable, gradual flow of income. It is not difficult to protect the landlord from inflation by using fair clauses. And the occupier has the prospect of using premises without being compelled to buy or to construct them. The reason why it is not very popular for firms to rent industrial property is financing. A tenant cannot offer his property as security for the mortgage he needs. However, conditions are changing in that respect. Increasingly, bigger firms which do not need mortgages are renting new premises for new branches. And since the German banks are not eager to lend money for capital ventures, newcomers often have to rent at first. Ownership is too burdensome and too much of a financial risk when starting a business. However, to some extent this is also a question of attitudes: businessmen like to show off their own premises.

Immobilienleasing (leasing)

A special form of tenancy, and one that is increasingly popular, is *Immobilienleasing*. In Germany the term 'leasing' is used when it is not a simple *Mietvertrag* or *Pachtvertrag* that is agreed. The owner, who often buys for this purpose, lets property to be used for a fixed sum comprising the price of the property as well as all cost, interest etc, payable in instalments. The occupier is fully responsible for the property. Usually a specified period for the leasing is arranged, often with an option to continue later on or to buy the property. Often a company will sell its property in order to lease it back after conversion to improve its condition; leasing is not only a kind of tenure, it is also a service. *Immobilienleasing* is usual only for new development or for totally refurbished buildings.

The occupation of property does not only entail compliance with the rules governing the relations between landlord and tenant, or observing the rights that ownership confers on the owner. Other legislation has to be observed by the occupier. In addition to planning legislation, building regulations, fire regulations, rules for the protection of employees and, last but not least, environmental legislation, are important. The *Bundesimmissionsschutzgesetz* (BImSchG, 1990 – Federal Act for protection against pollution), which sets up rules for clean air and noise, is very important. The Water Acts of the Federal Government as well as of the *Länder*, and Acts on waste disposal, have to be complied with. These legal requirements generate costs, but they also ensure a good environment and perhaps even a good image for some places.

3.1.2 Planning legislation

In Germany, with its federal constitution, there are three tiers of government:

1. the national, or federal, level;
2. the *Länder* (states); and
3. the local authorities, the *Gemeinden* (towns, regions) and *Kreise* (counties, districts).

The law defines strictly which tier of government has the legislative power for given matters.

The legislative sovereignty for the economy, for housing and for land law, including plans, is held by the Federal Government. For *Raumordnung* (spatial policy) the Federal Government can only set up principles as a binding framework for the *Länder*. The latter formulate rules and aims for the whole area of their *Land (Landesplanung)* or for parts of the *Land (Regionalplanung)*. The municipalities are responsible for detailed land-use planning.

In 1986 the Federal Government enacted the *Baugesetzbuch* (BauGB – Planning Code), governing detailed land-use planning. This is a consolidating Act, incorporating in one statute the previous legislation on town and country planning and urban renewal. It is supplemented only by the *Baunutzungsverordnung* (BauNVO, 1990 – Land Use Ordinance). The *Baunutzungsverordnung* is based on the *Baugesetzbuch*. It is normal legislative practice in Germany to incorporate only the main rules in major Acts, with the government or a minister being empowered to issue Ordinances to specify necessary details. The *Baugesetzbuch*, for example, allows the municipalities to determine in a local plan the land use and density and form of development permitted on a site. These rules can only be operational because the *Baunutzungsverordnung* specifies exactly what kind of land use zones (*Baugebiete*) can be authorized in a *Bebauungsplan* and what form of construction can be authorized within each land-use zone.

In addition, since the summer of 1990, the *Wohnungsbauerleichterungsgesetz* (WoBauErlG, 1990 – Act for facilitating the construction of housing), and some special rules in force only in the new *Länder* in the eastern part of the Federation, have to be taken into account.

Planning policy

Since trade and industry are important in a country with limited natural resources, policies for the creation of new industrial property have been developed at all levels of government. The Federal Government has the objective of creating living conditions in the whole country which are, if not equal, then at least of the same value. This principle is laid down in the *Bundesraumordnungsgesetz* (ROG, 1989 – Federal Act on Spatial Policy). This Act, together with the *Bundesraumordnungsprogramm* (Federal Program for Spatial Policy), emphasizes the necessity of a supply of jobs in all parts of the country. Industrial development is to be fostered particularly in those parts of the country which are below the average standard of the nation.

It is up to the *Länder* to formulate objectives for certain areas and regions. In *Landesentwicklungsplänen* or *Landesentwicklungsprogrammen* (plan or programme for the development of the state) they present aims for sectors of environmental and land-use planning, designating, for example, areas where industrial development is to be promoted and supported, or areas which should be developed for recreation or for housing. Though plans for the whole *Land (Landesplan)* or for part of a *Land (Regionalplan)* do not only formulate goals but may contain explicit land-use planning, and though development in small villages may not be desirable, there are no real obstacles for the development of industrial property if the demand for such property arises.

Since *Landesentwicklungspläne* are passed as Acts of the State legislature, these plans and principles are more than a policy; they are part of the legal framework and are binding for the municipalities, who often use their planning powers in accordance with these rules to respond to the demand of the market and to the needs of the developers if the intended development makes sense also for the municipality, creating jobs and broadening the tax base.

Most of the *Länder* have rules in their own *Landesplanungsgesetz* (State Planning Acts) about a *Raumordnungsverfahren*. This is a special procedure to be carried out when some bigger development is being prepared and planned: e.g. a new business park, a large retail centre, or a new motorway. The procedure has the objective of examining whether the development in question is compatible with the principles of regional policy. Deviations from these principles can result in a scheme being prohibited.

The legal framework for land-use planning

The competence and the obligation for binding land-use planning rests with the *Gemeinden* (municipalities). For every municipality a *Flächennutzungsplan* (land-use structure plan) has to be drawn up. A *Flächennutzungsplan* shows the essentials of the intended land use for the whole area of the village, town or city. It is binding for other planning authorities, if these participated in the process of drawing up the *Flächennutzungsplan*. Its main purpose is to prepare the *Bebauungsplan* (local plan), which is binding for everybody; a *Bebauungsplan* has to be derived from the *Flächennutzungsplan* and must not depart from it. Land use determined in a *Bebauungsplan* is permitted and authorized for everybody; any other land use is not allowed. Everybody can learn what kind of development will be allowed if the *Bebauungsplan* is looked at, perhaps supported by reading the *Baunutzungsverordnung*. There is no discretionary power of the planning authorities. The binding force of the *Bebauungsplan* is possible because the process to establish a *Bebauungsplan* is governed by the principle of *Abwägung* (weighing)

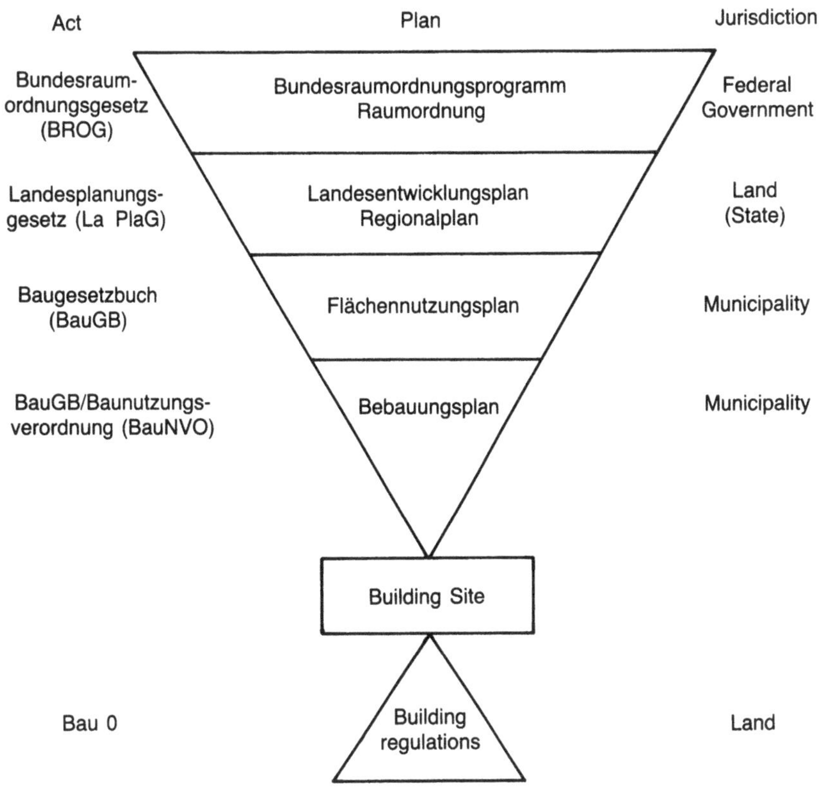

Figure 3.1 The planning hierarchy in Germany.

all conflicting interests fairly and correctly. Every *Bebauungsplan* is based on an environmental assessment according to EEC Directive EEC/85/337.

Figure 3.1 shows the planning hierarchy in Germany.

Development control

Planning permission is necessary for all *Vorhaben* (proposed development) to construct or alter buildings or to change the use of buildings. Under the laws of the *Länder* building permission is also required. Both permissions are given together in one act of the administration, the *Baugenehmigung* (permission to construct). In the following only planning law is explained; building regulations, e.g. for fire protection, or the rules for the protection of monuments, are beyond the scope of this chapter.

Applications for the *Baugenehmigung* are submitted to the municipality, which passes the application to the building authority which is working in contact with the planning authority for the area in which the proposed development is to be carried out. For industrial development there is no period of time in which an application must be determined. Often the determination takes months.

If the application is made for development within an area for which a 'qualified *Bebauungsplan*' (local plan) is in existence, the determination of the application is more or less a formal administrative procedure as far as planning law is concerned. A 'qualified *Bebauungsplan*' is a local plan which contains at least rules about the prospective land use (for housing, for mixed use, for industry, for recreation) according to the categories of the *Baunutzungsverordnung* (Land Use Ordinance), prescriptions about the measure of use allowed (e.g. how much floor space), the spaces to be built on, and finally the spaces for traffic. There is no discussion about the legality of development, as long as the proposed development does not depart from the *Bebauungsplan*. However, the application can only be favourably determined if it is guaranteed that the area will be serviced (roads, sewage etc.) by the time the planned buildings will be ready.

If the proposed development departs from the *Bebauungsplan* an exception can be applied for, but only if such exceptions are contained in the *Bebauungsplan*. It is also possible to apply for a dispensation from the rules of the *Bebauungsplan*, e.g. to build more floorspace than generally allowed. A dispensation is usually only possible if it is for the public benefit. And, of course, neighbours must not be touched.

Development may be authorized if the city council has decided to draw up a *Bebauungsplan* but the formal procedure has not yet been finished. If an application is in compliance with the future *Bebauungsplan*, and if the public has already been heard about the plan, the application may be granted.

In built-up areas with a continuity of buildings but without a *Bebauungsplan*, proposed development may be authorized if the building will fit into the existing built environment. Whether a building will fit in is judged by the land use in the surroundings and the type of coverage. In some municipalities up to 50% of all planning applications have to be determined in this way. Dispensations from this requirement are possible, but again only for the public benefit. Often this rule is applicable for industrial development, for instance if derelict land is being recycled in old industrialized areas. The possibility for dispensation is a direct indication of the policy of encouraging industrial development everywhere.

In outer areas, where no *Bebauungsplan* is in existence and without a continuity of buildings, development is not altogether impossible but it is not easy to get a favourable decision on an application for development. On green-field sites only agricultural development and development for public amenities are possible without legal difficulties. Only development which would disturb the neighbourhood in a built-up area can be allowed in the *Aussenbereich* (outer area). Almost always industrial development will disturb the public interest even if it is planned in the outer area. If this is judged to be the case, it will not be allowed.

In the five eastern *Länder* of Germany there is one more opportunity for development, particularly industrial development. The planning system in these *Länder* was quite different before 1990. Since it is not possible to draw up all the local plans necessary for a quick economic and social recovery of these areas, development can be permitted not only according to the BauGB, but also under § 55 *Bauzulassungsverordnung* (BauZVO, 1990 – Ordinance for the permission of buildings of the former GDR). Planning permission may be granted without a local plan if the development is urgent for the creation or the safeguarding of jobs, for housing or for infrastructure, and if a developer commits himself to carry out and finance a plan for special measures to support the development, including servicing of the area. The municipality has to determine the plan submitted by the developer, and the development itself is authorized by communal statute, like a *Bebauungsplan*. The provision is intended to shorten the lengthy procedure a local plan needs before it is applicable.

If somebody only needs to know whether a proposed development is legal without having drawn up all the plans for construction, he may apply not for a *Baugenehmigung* (permission to construct), but make a *Bauvoranfrage* (preliminary application), which may be restricted to certain problems of the development, and also to the legality of development under planning law. The possibility of a preliminary determination is often used and is very advisable, particularly in areas without a local plan. It gives security for a project at smaller cost than a complete application to construct.

If planning permission (as part of a preliminary determination or a permission to construct) is denied, the applicant may raise objections which are decided in a pre-trial proceeding by a higher authority. If the applicant is not satisfied with the decision he may go to court, probably three times: he can sue the municipality in the administrative court, he can lodge an appeal with the Higher Administrative Court, and he can lodge an appeal on points of law with the Federal Administrative Court.

Re-plotting of land – land assembly

Bebauungspläne are made for implementation. If the plots in the area of a local plan are small it is hardly possible to use the land properly as the local plan offers or calls for. Only if the existing boundaries are changed can the land be used according to the local plan. That is why the *Baugesetzbuch* offers a formal procedure for re-plotting of land and a method of land assembly. It has to be used if land ownership, as a consequence of the laws of inheritance, is widespread and the individual sites are unfavourably shaped.

The process of *Umlegung* (re-plotting of land, or land re-adjustment) facilitates the implementation of the local plan: every landowner has to contribute, in proportion to the size or value of his plot, to the land necessary for local streets and similar amenities. Instead of his old plot he receives a favourably shaped site which can be used for construction in accordance with the local plan. If the new land, although smaller than the old site because of the deduction for streets etc, is more valuable than the old site, the difference has to be paid to the municipality (or the municipality gets some more land).

Figure 3.2. shows a small area, zoned for industrial development before and after the process of re-plotting. The old and the new sites of some owners can be seen. The *Umlegung* in this area came to an end in 1990. In some parts of Germany the procedure of *Umlegung* is very common and, for green-field development, the usual way to develop industrial zones. *Umlegungen* can also be used in the rehabilitation process for neighbourhoods which are run down, and for industrial property.

Another possibility for land assembly is the *Städtebauliche Entwicklungsmassnahme* (Urbanistic Development Measure): not very often used in the past, abolished in 1987, and re-introduced in 1990. If a municipality has the idea to develop a large area for housing or for industry, the area may be determined by statute. The consequence of this is that the municipality has to buy all land in the area, if necessary by compulsory purchase for a price not including the influence of the expectation of the development measure. Having made the land suitable for building purposes the municipality has to sell the land (privatization) for its market value. There are examples of such areas where mainly smaller industrial property will be built.

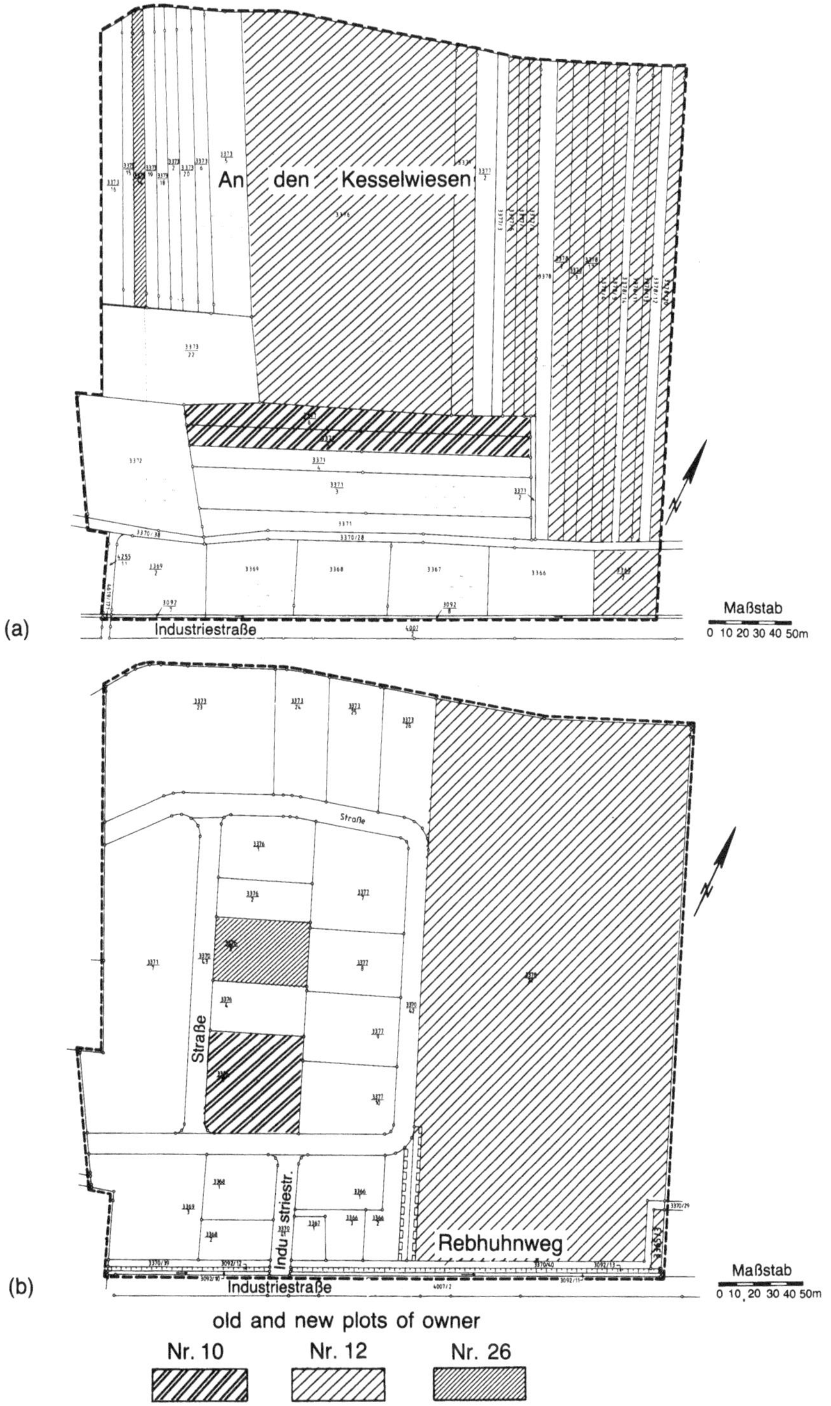

Figure 3.2 Umleng – re-plotting of land for industrial use. (a) Before re-plotting and (b) after re-plotting.

Originally Urbanistic Development Measures were always applied to green-field development. Today this instrument can also be used in inner-city areas for the recycling of derelict land and its re-use for industry.

Grundbuch – Land Register and local valuation committees

Another part of the legal framework for civil law is the *Grundbuchordnung* (GBO, 1935 – Land Register). This is a reliable record of all real property and its owners. There are no doubts about the title to a property, at least in the area of the former Federal Republic. This greatly facilitates the operation of the land and property market.

Property valuation is regulated in the *Baugesetzbuch* (Planning Code). It is therefore part of planning law. Local valuation committees collect information about land and property values and provide information about necessary data for valuation purposes, as indices of land prices, the rates of return for certain land uses on the local market, coefficients of conversion for prices for different land uses etc. The material and figures from the valuation committees are very reliable, since all contracts and deeds with which land is conveyed from one person to another have to be sent to the committee. It is advisable to use this source of information, at least for a first overall survey of land and property prices.

3.1.3 Taxation

The German *Steuerrecht* (tax law) has a significant influence on the industrial property market. Very often the taxation system decisively affects the buying, selling or renting of properties, for example by the possibility of saving taxes. It is generally held that the German property market is favoured by the tax system. Furthermore, the municipalities have a substantial influence in the taxation of properties.

Three main areas of taxation important for the functioning of the German industrial property market can be differentiated:

1. taxation for ownership (property tax);
2. taxation for selling or buying industrial properties;
3. taxation of the rent for industrial properties.

Ownership (property tax)

As mentioned above, ownership is the most common form of tenure of industrial property in Germany. Nevertheless the tax on the owner of properties, called *Grundsteuer* (GrStG, 1973 & 1976 – Property tax), is not very high; it indirectly promotes freehold ownership.

The *Grundsteuer* has to be paid by all owners of land, whether built up or not, and also by holders of hereditary leaseholds (*Erbbaurecht*, see above). The base for taxation is the *Einheitswert* (standard value of property) according to the *Bewertungsgesetz* (BewG, 1985 & 1987 – Valuation Act). Certainly the *Einheitwert* is different from the *Verkehrswert* (market value), because the *Einheitswert* is based on the value conditions in 1964 plus 40% of the real market value. In this way taxation doesn't limit the occupation or even the hoarding of land.

The municipality has the sovereignty over the *Grundsteuer*. Furthermore it is able and allowed to vary the tax rate by a special *Hebesatz* (leverage factor), typically 100–500%. Although the *Hebesatz* is a fiscal instrument to improve the communal finance conditions, it sometimes becomes an instrument to settle new firms by using lower *Hebesätze* than the community in the neighbourhood. In this way the *Grundsteuer* becomes part of the municipality's economic development policy.

The *Vermögenssteuer* (VStG, 1985 – Wealth tax) is also important for the owner. The basis of calculation is the whole wealth of a private person or of an enterprise. It includes the wealth in land, the wealth of business or other wealth. The tax rate is 0.5% for private persons and 0.6% for corporations (e.g. companies), although special *Steuerfreibeträge* (tax exemptions) exist. In most cases the owner of a property is allowed to subtract all mortgages on his property from the base wealth tax. So this tax is becoming less and less predominant.

The *Gewerbesteuer* (GewStG, 1984 & 1988 – Trade and business tax) is another important tax which must be mentioned in connection with the industrial property market. The basis of taxation is the *Gewerbeertrag* (income of trade) and the *Gewerbekapital* (capital of trade). The *Gewerbeertrag* is the determined profit from the business. The *Gewerbekapital* is the *Einheitswert* (standard value) of the firm. In Germany each firm or company has to pay it, if the main income derives from business. The *Gewerbesteuer* is the most important finance base for all German municipalities. The local authorities are allowed to fix and to vary the tax rate by a special *Hebesatz*. For this reason the municipalities are very interested in increasing the number of firms in their town area. Competition between communities is normally the consequence.

Although the real burden of costs is not too high, the tax rate of the *Gewerbesteuer* often influences the whole 'economic atmosphere' in a community.

Selling or buying industrial properties

The transfer of ownership from one owner to another is taxed by the *Grunderwerbssteuer* (GrEStG, 1982 – tax for buying and selling properties). Normally the tax rate amounts to 2% of the price paid to purchase the property. If there are no fixed prices the basis of taxation depends on the *Einheitswert* (standard value of property). Normally the buyer is liable for the tax, although according to the *Grunderwerbssteuergesetz* (GrEStG, 1982 – Act for taxation of the buying and selling of property) the buyer and the seller are both responsible for the payment.

The *Grunderwerbssteuer* nearly always has to be paid as a requirement for the entry into the *Grundbuch* (land register) but for closed real estate investment funds, for example, different and complicated privileges such as tax exemptions or tax cuts are available.

The gains from buying and selling properties are also liable to *Einkommenssteuer* (income tax) and *Gewerbesteuer* (trade and business tax), if the sale is a business transaction; the private sale of properties is normally tax-free.

Another important tax for the functioning of the real industrial property market is the so-called '*Spekulationssteuer*' (speculation tax). The seller has to pay this if he sells the property at a higher price within two years after having bought it. Unfortunately the *Spekulationssteuer* is a limited instrument for preventing excessive speculation in the market. Furthermore in practice it is difficult for finance administrators to distinguish between private and professional transactions.

Despite their limitations, the *Grunderwerbssteuer* and the *Spekulationssteuer* can discourage the owners of properties from frequent selling.

Rental

There is no specific tax for rental properties in Germany, but other taxes influence the rental market. The proceeds from renting and leasing properties have to be added to the whole income (private and company income), so that they are also liable to the *Einkommenssteuer* (income tax). If renting and leasing of properties is the main business, the companies have to pay also *Gewerbesteuer* (trade and business tax).

3.1.4 Subvention

The German economy is considerably enhanced by various programmes of subvention. It is appropriate to subdivide the subsidy system into subventions for the entrepreneur and for the municipality.

The municipalities can usually ask for subventions directly in the form of grants from the *Städtebauförderungsmittel* (subsidy for urban improvement) programme, awarded each year by the *Länder*. In this case the municipality is allowed to spend the money for the developing of new industrial areas and for the assembly of new building land.

Entrepreneurs have a wider choice of finance from different public authorities. All relevant and available subsidies are collected once a year by the special magazine *Zeitschrift für das gesamte Kreditwesen*. This provides an excellent overview of the predetermination for a specific purpose, the conditions of awards, and the method of application. For example, the Federal State awards subsidies in different forms (grants, bonus payments, loans, sureties). The subsidies are subdivided in numerous special programs for different purposes(eg *ERP-Sondervermögen* or the *Gemeinschaftsaufgabe: Verbesserung der regionalen Wirtschaftsstruktur)*. These funds are administrated and awarded by the banks *Kreditanstalt für Wiederaufbau* and the German *Ausgleichsbank*.

At the State level there are supplementary finance programmes (grants, bonus payments, loans, sureties) to support policies with special purposes such as regional development, or new technology. In addition it is likely that there will be some concentration of subsidies on the new states in the eastern part of Germany.

3.2 MARKET

There is not really a homogeneous industrial property market in Germany. In addition to the various market sectors (warehouses, production buildings, service institutions, etc.), there are wide regional differences.

One can no longer talk about real and free-market conditions in structurally weak or peripheral regions (rural areas, old industrialized areas). The considerable influence exerted by the public sector, especially on industrial building land, creates an artificial land price. This can be so low that the price of the real estate has hardly any influence on the investor's decision. These conditions are normal in many parts of Germany. With the exception of a few prosperous agglomerations this can be regarded as a more usual land market for industrial use.

Only in those cities and regions of Germany which show significant economic growth can one speak of a property investment market. However, even in economically weak areas there is a user market for service, distribution and production buildings, but this is hardly developed and is of minor importance.

Each of the case studies in section 3.3 is typical of one of the market sectors.

3.2.1 User market

It is mainly the public sector, and usually the community, which decisively influences the industrial property market. With planning instruments and massive assistance with finance, organization and advice, the municipality tries to attract firms and to enlarge the supply of industrial property by:

1. providing cheap building land (reduced in price);
2. intensified promotion;
3. creation of infrastructure in advance;
4. creating its own stock of buildings (but only very rarely as yet).

Spectacular intercommunal shifts are rare: instead there are intracommunal shiftings of firms which already exist locally. The aim of the municipality is not so much a monetary orientation for yield as the creation of new jobs and a strengthening of the structural weakness of the economy.

The market conditions described above can generally be found in the following types of area in Germany:

1. areas industrialized to a low degree, or marginal zones of agglomerations;
2. old industrialized areas;
3. rural-peripheral areas.

A low degree of acceptance concerning the rental and leasing of these properties is typical for this market. The combination of acquisition of land with the subsequent establishing of buildings and the use by only one single entrepreneur or company is widespread in Germany and is still dominant. Often industrial property is financed in a conventional way: 60% credit on landed property, 20% personal credit and 20% entrepreneur's own resources.

Used premises are normally sold together with the land as a unit. However, the *Erbbaurecht* (hereditary leasehold) is also applied.

It is almost always the local authority which draws up any plan, usually a *Bebauungsplan*. The public sector is responsible for servicing and re-plotting of the building land; streets remain public property.

Normally the municipality tries to keep a large reserve of building land in different locations, for different sectors and in various conditions of development. Generally the market behaviour can be regarded as an extensive supply policy. In this way the local authority becomes a *Zwischenerwerber* (intermediate acquirer of land) by buying and developing the land and selling it to different applicants.

With its legal instruments (planning, re-plotting, right of pre-emption) the municipality is able to provide various plots and to offer them to interested and interesting companies. Sometimes the municipalities promote their industrial building land by information papers or by nationwide newspaper advertisements. Meanwhile the promotion of industrial building land and the acquisition of new firms are increasingly the responsibility of the municipality's *Wirtschaftsförderungsamt* (Economic Assistance Board).

In recent years new *Öffentliche Wirtschaftsförderungsgesellschaften* (public economic assistance agencies) have been growing rapidly, especially in large cities. They have the same aims as the economic assistance boards but are more flexible and not over-involved in the communal administration system.

If an entrepreneur is interested in buying a plot of land he normally enquires directly from the municipalities. It is not common to hire a broker or a real estate agency for searching. Privately owned plots or used premises are often advertised by the owner in a newspaper advertisement. If an applicant wants to ensure that he is being asked a fair market price, he can get information from the independent *Gutachterausschüsse* (local valuation committees) described above.

The municipality is responsible for developing new industrial areas, and for bearing the development costs. Although, according to the BauGB, the owners have to pay for servicing their plots, there are special subsidies from the States and the *Bund* (Federal government) for the purpose. Usually the public space (streets, green spaces) remains in the ownership of the municipality.

Development gains are theoretically possible, but because of the competition between municipalities, the offered prices are too low and most municipalities therefore make a loss.

The creation of a building stock by the municipality is not common in Germany, although the market in this kind of industrial property has increased in recent years. Some municipalities (e.g. Köln) have built such property, of a simple form of construction, for renting. Apart from this, factory building by government does not exist in Germany. Nevertheless, the founders of new firms, who normally have limited financial resources, often have a particular need of such premises.

There is no government control of land prices and rents, or of rents for industrial buildings.

3.2.2 Industrial property investment market

In growing areas, where economic efficiency seems to be most secure (i.e. Frankfurt/Main, Munich, Hamburg, Düsseldorf), an independent industrial property investment market has developed by the initiative of private project-developers. Increasingly, foreign investors from such countries as Japan, Sweden, USA, Great Britain and the Netherlands, are investing capital. German individual and institutional investors are active too. The aim is to invest capital in industrial properties expecting yields from selling, renting or leasing them.

High-quality industrial parks, in locations which are attractive and conveniently situated and favourable for traffic, constitute an investment category of particular interest. This form of development is characterized by the creation of a large number of different industrial buildings by developers who normally don't use the buildings themselves.

The behaviour of private project-developers, and their cooperation with other participants in the industrial property investment market, varies with different working methods and different market positions. Furthermore, the course of events on the market is complicated, because there are few legally binding rules and the competences of the different participants are continually changing.

The rest of this section describes the common types of operation on the industrial property investment market from the point of view of private project-developers.

In recent years most private project developers have operated on the market according to a particular pattern. First, the developing companies normally search intensively for suitable locations to develop profitable industrial properties. If the locations aren't in the same region the companies commission specialized agents to locate likely places. At present it is not common to ask the municipality for suitable locations, although an increasing number of towns in Germany are co-operating with private project-developers in this way.

If a suitable and profitable location is identified, the developer makes a *Bauvoranfrage* (preliminary application). Given a positive statement from the municipality the project developer drafts a conception of use, ascertains the different building costs and calculates the possible gains from rental. Models for financing the project are considered.

Usually the developer obtains from the landowner a *notarielles Verkaufsangebot* (offer to sell) attested by a notary, to retain the chance to decide later which client is buying the plots. When the client is known the developer normally arranges a contract with a withdrawal clause, if the municipality does not grant definite building permission or if the profitability cannot be ensured.

Often a local broker, hired by the developer, concludes contracts with the later tenants of the properties, although up to this point no building has been erected. Nowadays, developers increasingly operate without creating their own stock of buildings, as the banks specializing in mortgages are not willing to finance any project without a guaranteed gain.

The developer contacts the later main tenants to ask them for their individual interests to incorporate them into the planning process. After this the developer offers this 'complete package': design concept, planning permission, fixed leases and fixed contract for sale of land directly to interested national and international companies. Sometimes he advertises in nationwide magazines.

The developer realizes a certain yield, the so-called 'developer profit'. The average profit, about 10% of the whole property value, is the difference between the costs of the whole construction (planning and building costs) and the agreed price of sale.

After the contract is signed the investment capital companies commission a building contractor to build the production space as well as the roads and access.

The management of the properties depends on their size. Sometimes the management of larger properties (e.g. large real-estate parks) is organized by special management companies, which are responsible for the acquisition of new tenants, for collecting the rent or lease, and for the whole condition of the property.

In addition to the form described above, sometimes private project-developers build and finance the projects (i.e. a real estate park) themselves and consider the properties as their own capital investment for the future. In this situation the centre of their interest is not quick profit but rather long-term investment. The developer's behaviour in the phase of beginning the project is similar to the previous case. He usually finances the project with his own and outside capital. He gains his profit by rental income from the properties.

Normally a development company possesses only a few industrial properties (e.g. an industrial real estate park), because maintaining their own department to manage the properties is not profitable. The main benefit for the development company in managing their own property is the opportunity to demonstrate its features to interested customers.

The creation of stock buildings by developers or other entrepreneurs is not common. The financial risks are higher, and so the main development companies are not very interested.

The influence and control of public authorities is of minor importance as far as project developments are concerned. The developer is responsible for planning and sometimes for developing the land too; streets often remain as private properties and the official approval frequently is given without a *Bebauungsplan* (local plan) according to § 34 BauGB.

The industrial property market, as a market in the sense of an 'investment market', will gain importance with the advent of the SEA and the expected increasing concentration of economic growth in the agglomerations in the south of Germany and along the Rhine. It will also become important for the new east German *Bundesländer* (states).

3.3 CASE STUDIES

These case studies try to point out present and future trends, which can be extrapolated to other areas in Germany. Although the distance between the examples presented (Dortmund and Düsseldorf) is only 80 km, the market conditions are completely different.

3.3.1 Dortmund technology park

The Dortmund Technology Park (Plates 1–4), is a very successful example of the establishment of new industries in Germany. Different private and public partners (Federation and state, regional government, municipality, chamber of commerce) achieved the aim of attracting high-quality industry to a structurally weak region. Dortmund is a kind of prototype for subsequent projects in Germany (e.g. Dresden Technology Park). The high qualitative situation of this park is still not the rule, but it shows the general trend to higher quality of industrial objects.

Dortmund, a city with around 600 000 inhabitants situated in the eastern part of the Ruhr agglomeration, has a technology park located on the outskirts of the city next to the University (Figure 3.3). Though an *Oberzentrum* (central city), Dortmund in contrast to Düsseldorf is an old-industrialized and structurally weak area. In former times Dortmund was dominated by the coal, iron and steel industries. In recent

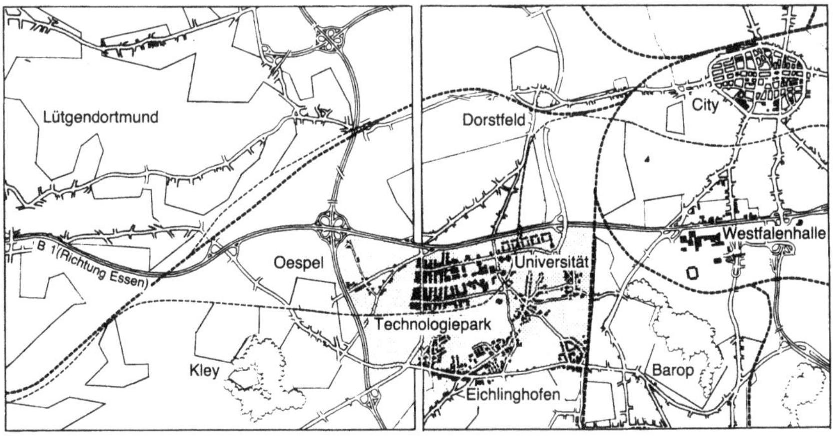

Figure 3.3 Location of Dortmund Technology Park.

years structural change has taken place accompanied by a high rate of unemployment. Dortmund finally accepted this change and tried to attract other industries.

At present the Technology Park is still in the growth phase. After its final completion it will comprise a total area of 11.5 ha with 80–100 firms and around 4000–5000 jobs. The park consists mainly of technology-orientated, medium-sized firms (chip production, material technology, environmental technology, etc.). The firms profit from the scientific orientation of the University. The exchange of staff, research results and development techniques is increasingly important.

The park was sited in an area which was in the possession of the city of Dortmund and was originally determined for the area of the University. Its traffic connections (e.g. direct motorway connection) are favourable. The urban and structural conception demands a high standard of quality, which clearly differs from the appearance of conventional industrial parks. Particular importance was attached to a standardized design of the public area, the facades, heights of the buildings, parking places (including multi-storey parking) and of the green area (figure 3.4). The essential instrument for planning control by the local authority, and its means of carrying out its aims, is a restrictive and detailed *Bebauungsplan* (Figure 3.5) and a *Gestaltungshandbuch* (handbook for design). These give interested firms good information as far as the site-purchase contract is concerned.

The design of the new park follows a strictly 'orthogonal' pattern, which forms single block structures consisting of 'site modules'. The normal site is 1900 m^2, and the smallest possible module 850 m^2. The

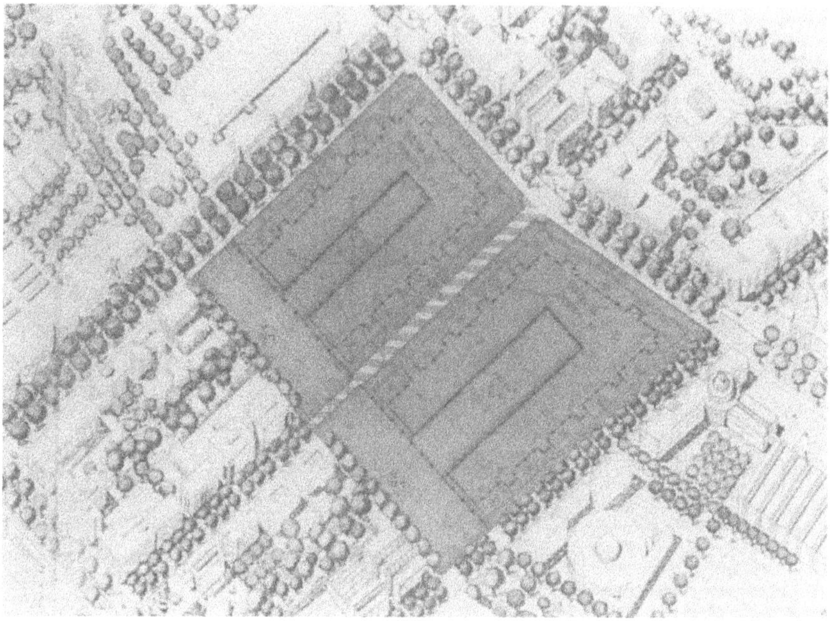

Figure 3.4 A typical block.

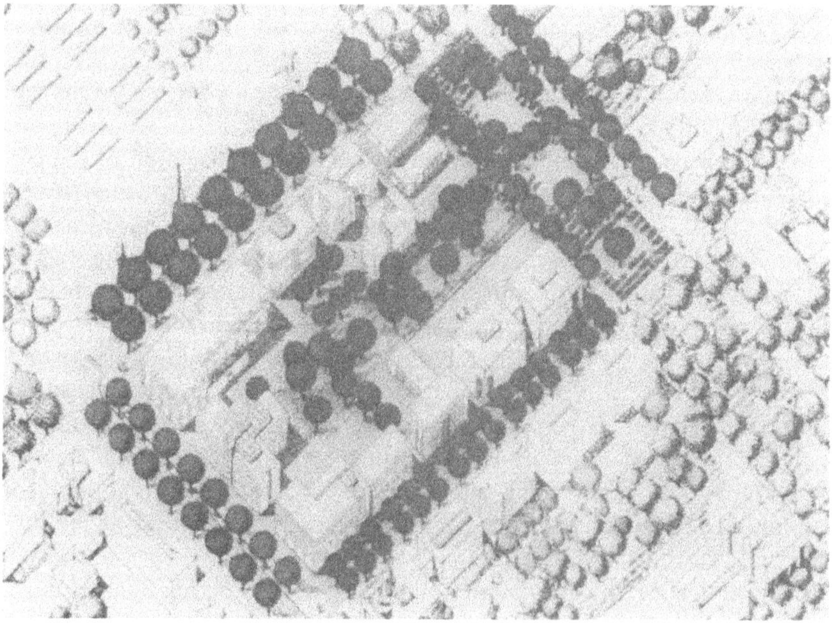

Figure 3.5 Local plan.

office and other service buildings with sight-lines to the streets have to be constructed with three floors, along a fixed *Baulinie* (building limiting line) and with red bricks only (Figure 3.5).

The planning and development, and the attracting of firms, have been undertaken entirely by municipal institutions. With this 'integrated planning concept', from the outset the developers followed consistent aims and design concepts in order to speed up the process of planning and implementation. This is a result of the new cooperation between various community institutions (town planning, economic assistance, green area office, environment office), with representatives of the State Government and of the responsible administration of the region of Arnsberg as well as representatives from the chambers of commerce and a consortium of banks.

A project group was formed, consisting of representatives of the groups mentioned above under the leadership of the *Oberstadtdirektor* (Director of the City Administration) from the city of Dortmund. Essential parts of the planning process included a *Bereichsentwicklungsplan* (special structure plan) for the area around the University, the realization of an urban design competition for the industrial park area, and the preparation of the *Bebauungsplan*.

All the development costs came from municipal resources and from the Ministry for Town and Country Planning, Housing and Traffic in North Rhine-Westphalia (MSWV). The aim of this project was the creation of new jobs and the hope of influencing other areas to accelerate structural change.

Interested firms are acquired by the *Wirtschaftsförderungsamt der Stadt Dortmund* (office of economic assistance of the city of Dortmund). The sites are sold at DM 130/m^2 to firms, with no charge for servicing (*erschliessungsbeitragsfrei*) and first planting (*kostenfreie Erstbepflanzung*). The development of the area was carried out by the community in advance and cost around DM 30 million. Roads and sewers stay in the possession of the municipality.

By special clauses in the purchase contract the Municipality maintains its influence on the area:

1. *Wiederkaufsrecht* – the right to re-buy the land if the buyer does not use it within a period of 10 years.
2. *Vorkaufsrecht* – the right of pre-emption;
3. *beschränkt persönliche Dienstbarkeiten* – restricted easements for the municipality, giving it the right to influence the design and architecture of new buildings (planning);

4. responsibilities for construction (planning) to ensure beginning of construction within four months after signing the contract, and *Baufertigstellung*, completion of construction within 12 months;
5. responsibilities for the design of the buildings according to the handbook for design, and for open areas according to a plan for green arrangement;
6. the right to approve of any *Nutzungsänderungen* (change of land use), *Weiterverkauf* (resale), renting, *Gestaltungsänderungen* (changes of design).

The park is a big success and the city of Dortmund is working on its extension.

3.3.2 Hansapark, Düsseldorf (plates 5–7)

This example represents the industrial property investment market and was chosen because of the dense concentration of industrial parks in Düsseldorf and in its surroundings. The park covers 8.4 ha and is located 6 km from the centre of Düsseldorf. Düsseldorf has around 600 000 inhabitants, and is one of the largest cities in the Rhine-Ruhr agglomeration. Many national and international companies have their head offices there, and a lot of companies control their production in the industrialized Ruhr area from there.

Düsseldorf, along with Munich and Frankfurt, has to be classified as a fast-growing metropolis. The city was spared the bigger structural changes of the old industrialized Ruhr agglomeration east of Düsseldorf. Furthermore, Düsseldorf has a number of administrative functions. It is the capital of North Rhine-Westphalia, has a large share in the service sector and many technology-orientated firms in the production sector (chemistry, microelectronics, environmental engineering). Düsseldorf has grown into one of the leading investment markets in Germany. In particular, more and more Japanese firms are settling there.

The park was established on derelict land, the site of a former iron foundry. Connections to existing traffic networks are very good because of the park's direct motorway connection and its nearness to the international airport. The integrated inner-area location, in contrast to a green-field project, constrains the design and land use concepts because it is much more limited by other uses in the surrounding areas. This applies to the design concepts of other industrial parks in Düsseldorf as well.

The planning concept, with looped road patterns, makes it possible

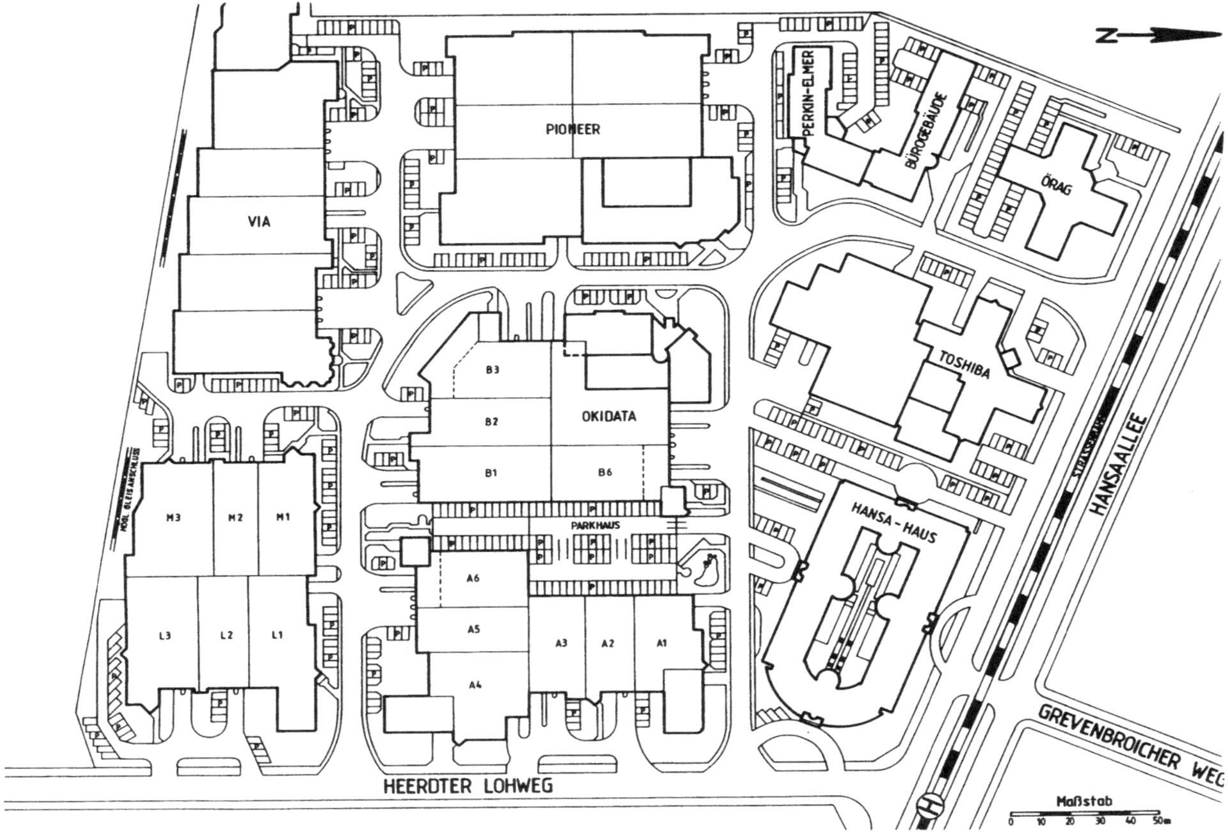

Figure 3.6 Ground map.

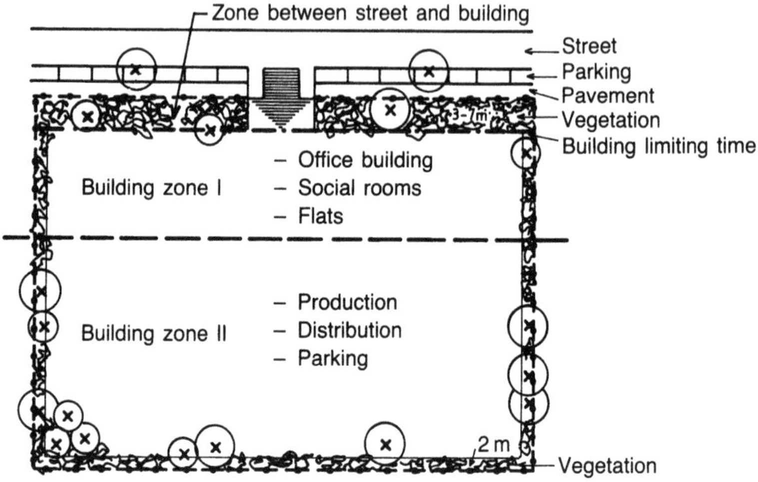

Figure 3.7 Principle of zones.

for all plots to be of comparable value insofar as their location is a variable. Almost all plots of land have two approach roads (Figures 3.6). The buildings are arranged according to the same general pattern (Figure 3.7): buildings and facilities of the service sector face the road whereas the production halls are in the background area. This gives an attractive and standardized impression of overall urban quality. There is adequate parking space in a central multi-storey car park and along the roads; the project developer is the owner of all common space (roads, green spaces). In comparison with public roads in industrial areas the cross-section of the streets (5–6 m) does not take much space. The building area covers only 50% of the site; 25% is for traffic and 25% for green areas. Therefore the building area must be regarded as comparatively small.

Planning began in 1982, with completion in 1989. A total of 33 firms have been established, mainly branches of computer and communications electronics firms, with an average of 45 employees per firm and 1500 employees in total. The average size ranges from around 1300 m^2 to 1800 m^2. The relation of office area to production area is nearly 50/50, which is comparatively high. The office buildings, with their attractive façades and prominent entrance areas, have two to four floors. Behind them, the storage and production halls (between 6 and 7 m high) are constructed according to a module system, allowing great flexibility to adjust to the needs of new tenants.

The whole industrial park was developed and arranged by a single private project-developer by way of creating large numbers of buildings. The developer did not offer the whole concept to investment companies, but instead rents all buildings and parking places to interested firms, so the park is a capital investment for the developer.

The rental price is fixed in advance by contract; however, price increases are not excluded. At the outset, the rentals for office space (DM 14/m^2) and for service and production space (DM 7.8/m^2) were not low; after the first price rise (after four years) the price for office space has reached DM 15–18/m^2, and for other space DM 8–9/m^2. The contracts with the firms, concerning the duration of the tenancy, vary. The first tenants could participate in the planning phase, and were therefore able to amend the design to meet their individual wishes.

The project developer administers and maintains the park. The recruitment of interested firms is carried out by professional agents, who try to achieve a mixture of activities, in co-ordination with the developer.

The public sector has had no influence either by planning or by subventions. Building permission was granted according to § 34 BauGB (within a built-up area with a continuity of building without a local plan).

3.4 DATA

There is no suitable information system that provides comparative data and figures specifically for the industrial property market in Germany. For example, uniform statistics are not kept for the interesting rental market, because this special market is not well developed and is only just beginning to grow. Another problem is to ensure the compatability of data. How can building quality and quality of location be compared?

As mentioned before the industrial property market has until now been mainly a market for industrial building land. During the last five years the turnover of building land has grown rapidly (Figure 3.8) and the trend continues. But there are great regional differences. In growing agglomeration areas pressure on the market is so high that there is a severe shortage of available land, causing price rises; whereas in the rural periphery of agglomerations, or in other rural regions, there is adequate supply and prices are low as a result of weak demand and massive public subsidies.

Figure 3.9 shows extremely varied prices for industrial land. Real market conditions are to be found in the main important growing areas like Munich, Frankfurt, Stuttgart and Düsseldorf. In Munich, for example, the rise in prices in one year (1988–1989) was nearly 38%.

	abs. 1980	%	abs. 1984	%	abs 1988	%
(a)	2.190	2.2	2.424	3.7	4.632	6.6
(b)	1.228	10.4	909	13.3	2.290	26.8

(a) Numbers of purchase (b) changed area in ha
(Source: Statistisches Bundesamt, Fachserie 17, Reihe 6 überarbeitet nach: Müller-Kleißler, R. und Rach, D.: Der Baulandmarkt in der Bundesrepublik Deutschland, in: Informationen zur Raumentwicklung Heft 6/7. 1989. 8. 402)

Figure 3.8 Turnover of building land for industrial land use in Germany absolute and in % of all purchases and of the area changed.

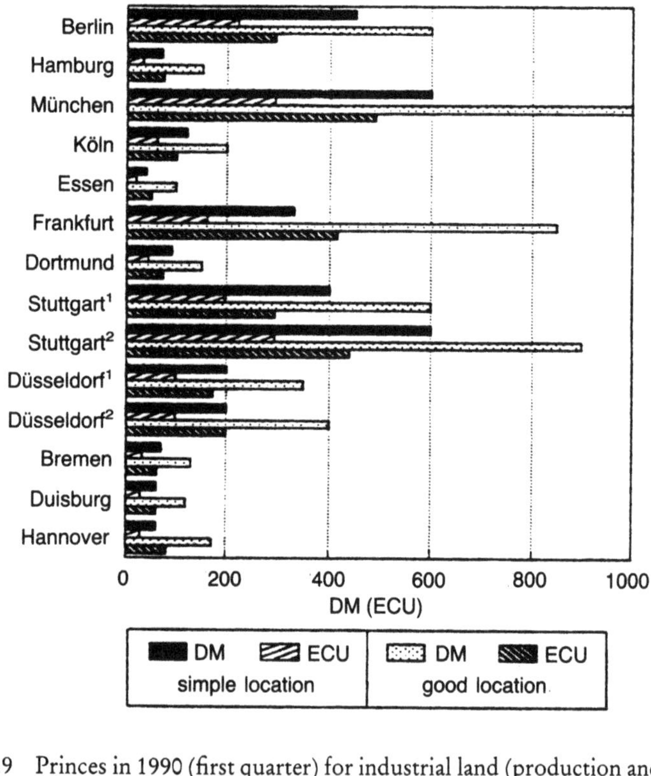

Figure 3.9 Princes in 1990 (first quarter) for industrial land (production and service) DM(ECU)/m³ (developed areas without subsidies). [1], lowest prices; [2], prime prices.
Source: Ring Deutscher Makler (RDM) e.V., Immobilien-Preisspiegel 1990.

More and more speculation is the consequence. However, national average prices (based on a comparison of municipalities) were not more than DM 40/m² in 1988. The projected demand for industrial building land amounts to nearly 1470 ha each year up to the year 2000.

The average plot size nowadays is around 3300 m^2. Normally new plots are not more than 5000 m^2. However, very small premises are not very common in Germany. The average size of the production and service space vary; for warehouses it is around 900 m^2 and for production areas around 1200 m^2.

Prices for renting industrial space differ greatly, and are only available for the main growing towns. Figure 3.10 shows prices for both mediocre and very well-located areas. According to Jones Lang Wootton (1990) rents are higher (Figure 3.11). In comparison to this, in Dortmund and Cologne, prices are between DM 4 and DM 8/m^2.

Building costs differ depending on the quality of the construction. Production space is normally in lightweight construction (concrete ceiling and simple flat roof), usually costing around DM 600–700/m^2, with around DM 100/m^2 to be added for a wholly concrete construction. For a high-quality standard of finish (e.g. by using red bricks), the costs come to DM 1000/m^2 and more.

City	rent (DM/m^3/month)	ECU
Düsseldorf	7.50– 8.50	3.7–4.1
Hamburg (only production space)	4.00– 8.00	2.0–3.9
Frankfurt	7.00–14.00	3.4–6.8
Stuttgart	6.00–10.00	2.9–4.9

(Source: Deutsche Immobilien-Partner, Hamburg, München, Stuttgart, 1989)

Figure 3.10 Lowest and prime rents for production – and service space in selected industrial estates 1989.

	Warehouse/Production space		Service space	
	DM	ECU	DM	ECU
Stuttgart	11.50	5.5	15.00	7.2
München	10.50	5.0	14.00	6.7
Frankfurt	10.50	5.0	14.00	6.7
Düsseldorf	10.00	4.8	11.40	5.5
Hamburg	9.50	4.6	12.00	5.8

The rents quoted relate to 'high-tech' accommodation – and show the top rent achievable for the best quality buildings in prime industrial locations

(Source: Jones Lang Wootton Consulting & Research, Market Report West Germany 1990)

The cost for service and office space amounts to DM 1000–1800/m^2, depending upon the quality standard. The average price is around DM 1200/m^2.

Figure 3.11 Prime rents for industrial space 1989/1990 per m^3 per month, ECU (Qualifying date May 1989).

REFERENCES AND FURTHER READING

Battis, U. (1987) *Öffentiches Baurecht und Raumordnungsrecht*, 2nd edn, Kohlhomme, Stuttgart.

Braam, W. (1987) *Städtplanung: Aufgabenbereiche, Planungsmethodik Rechtsgrundlagen*, Werner, Düsseldorf.

Bundesministerium für Raumordnung, Bauwesen und Städtebau (eds) (1986) *Städtebauliche Bericht: Umwelt und Gewerbe in der Städtebaupolitik*, Sonderheft, Schriftenreihe Staedtebauliche Forschung, Bonn.

Dieterich, H. (1985) *Umwidmung brachliegender Gewerbe – und Verkehrsflächen*, Schriftenreihe Städebauliche Forschung, Bundesministerium für Raumordnung, Bauwesen und Städtebau (eds), Bonn.

Dieterich, H. (1990) *Baulandumlegung*, 2 edn, Beck, München.

Dose, N. and Drexler, A. (eds) (1988) *Technologieparks: Voraussetzungen, Bestandsaufnahme und Kritik*, Westdeutscher Verlag, Opladen.

Erbguth, W. (1989) *Bauplanungsrecht*, Beck, München.

Ernst, W. Zinkhan, W. and Bielenberg, W. (1990) *BauGB Kommentar*, Stand 43 Lieferung, Loseblattsammlung, Beck, München.

Falk, B. (ed) (1985) *Immobilien – Handbuch, Wirtschaft, Recht, Bewertung*, Kohlhammer, Stuttgart, Berlin, Köln, Mainz.

Falk, B. (ed) (1987) *Gewerbe – Immobilien* Lech, Landsberg.

Finkelburg, K. and Orthoff, K.-M. (1990) *Öffentiches Baurecht*, Band I: Bauplanungsrecht, 2nd edn. Beck, München.

Gerlach, Heinz, *et al.* (1988) *Die Gewerbeimmobilie als Kapitalanlage*, Freiburg i B.

Hennicke, M. and Fengler, H. (1985) *Industrie- und Gewerbeparks als Instrument der kommunalen Wirtschaftsförderung*, Schriftenreihe zur Mittelstandsforschung Nr 4, Stuttgart.

Heuer, B. (1990) *Erfolgreiches Vermarkten von Gewerbeimmobilien*, Loseblattsammlung, Weka.

Heuer, H. (1985) *Instrumente kommunaler Gewerbepolitik*, Kohlhammer, Stuttgart.

Hüttenrauch, Christian, Dzelzkains, Arnolds (1987) *Markt- und Preisverhältnisse für Gewerbe- und Industriegrundstücke*, Der Städtetag 11. Köln.

Hüttermann, A. (1985) *Industrieparks. Attraktive industrielle Standortgemeindschaften*, Wiesbaden, Steiner, Stuttgart.

Jones Lang Wootton GmbH (1990) *Der Büroflächenmarkt in der BRD*, Markt Bericht BRD 1990. Jones Lang Wootton, Frankfurt a.M.

Palandt, O. (1990) *Bürgerliches gesetzbuch – Kommentar*, 49th edn. Beck, München.
Rössler, R. Langen, J., Simon, J. and Kleiber, W. (1990) *Schlätzung und Ermittlung von Grundstuckswerten*, 6th edn. Campus Verlag, Frankfurt.
Schmidt-Eichstaedt, G.(1987) *Einführung in das neue Städtebaurecht – Ein Handbuch*, Kohlhammer, Stuttgart, Berlin, Mainz.
Schrumpf, H. (1984) *Technologieparks als Instrument kommunaler Wirtschaftsförderung*, Rufis, Bochum.
Troll, M. and Simon, J. (1986) *Wertermittlung bei Geschäfts- und Fabrikgrungstücken Handbuch*, 2nd edn, München.
Verlag C. H. Beck (ed) *Rechtshandbuch Immobilien, Band I: Bauen und Finanzieren; Band II: Vermitteln, Verkaufen, Vermarkten.*
Weirich, H.-A. (1985) *Grundstücksrecht*, Beck, München.

Plates 1–4. Examples of industrial property in the Dortmund Technology Park.

Plates 5–7. Premises in Düsseldorf Hansa Park.

4

SPAIN

Enrique Calderon and Ignacio Espanol

4.1 GOVERNMENT AND PLANNING CONTEXT

This chapter aims to provide a basis for understanding the functioning of the industrial property market in Spain from the viewpoint of the planning process which legitimizes the provision and trading of this type of property.

As in other European countries, authorization to use land for specific forms of development or economic activity must be sought from the appropriate public authority, in accordance with pre-established land-use plans or pre-formulated public policies. In Spain, the drafting and approving of land-use plans falls primarily within the jurisdiction of municipal government, although other layers of the public administration, namely central (national) and regional, may also have some say in controlling and authorizing development. As for general policy formulation, central and regional administrations play the most important roles. To understand these mechanisms of plan and policy formulation is essential to meet the objectives of this book. Section 4.1.1 will be devoted to an analysis of the responsibilities of each of the different bodies which may influence authorization of development, in accordance with the constitutional principles and the prevailing national laws.

Once the plan has been approved, the subsequent control of development basically lies within the responsibilities of the municipal authorities. Detailed plans may be required in accordance with the overall specification of the municipal plans so as to establish a set of conditions for the relationship between the authorities and prospective developers. This relationship is fairly rigid, but this does not mean that there are not ways to modify land uses outside the complete procedures of plan revision.

Section 4.1.2 therefore deals with the detailed aspects of the management of municipal plans within which industrial development must proceed, as well as the options open to the prospective developer to bypass the plan specifications: in other words, to bypass the preset supply-demand relationship of land available for industrial uses embodied in the approved plan.

A final set of issues to be dealt with refers to the financial aspects linked to industrial development. These, in turn, involve the analysis of the different actors who may initiate development, be it public or private, and the financial options open to them; and of the classification of the Spanish territory according to the level of industrial aid available from national or supranational (i.e. EC) sources. This, of course, reflects not only the current situation in terms of development and resource availability, but is also an expression of the prevailing spatial policy for industrial development in the country. As in many other issues in the Spanish context these days, this spatial policy is a patchy overlap of formulations between the different levels of government affected. In section 4.1.3, therefore, financial issues are tackled including a classification of area incentives and the analysis of the different actors who take part in development.

Throughout the text a series of concepts will be introduced, as well as references to legislation and administrative by-laws, which will be listed in the appendix to this chapter.

4.1.1 The government of planning

To understand the options open to the developer in the industrial property market in Spain requires a knowledge of planning policy formulation, since it is within the specifications defined in the plans that authorization for development may be granted. As mentioned above, central and regional administrations can intervene in the final shape of a plan, although the main responsibility for the ultimate physical structure of the territory lies in the hands of the municipal authority.

The share of responsibilities within the various administrations involved in plan-making is established in the Spanish Constitution, approved on 6 December 1978. Our Magna Carta organizes the country into a quasi-federal system with 17 autonomous regions each endowed with a parliament and a government. This system falls short of that of the Federal Republic of Germany, being more similar to the Italian system, whose own constitution has no doubt influenced the Spanish one. As a first consequence of the coincidence of 17 regional plus one national legislative and executive powers, there is bound to be some overlap in all fields, and spatial planing and the management of the territory is no exception. The share of powers between national and regional administrations is established in Articles 148 and 149 of the Constitution. Article 148 bestows upon the autonomous communities several responsibilities linked to industrial development, including:

1. *ordenación del territorio, urbanismo y vivenda* (see below),
2. public works, for the interest of the regions, within their own territory;
3. roads and railway lines whose complete route runs within their territory;
4. the management of environmental protection;

5. economic development within their territory in accordance with national economic policy;
6. industrial promotion, within the national industrial policy, and a mixed system of industrial aids jointly managed by both administrations.

The meaning of the concept of *ordenación del territorio* has been the subject of debate amongst experts, its goals swinging from a purely economic activity aimed at fostering regional economic development to a more executive endeavour in which land uses and infrastructure provision are co-ordinated in order to direct population and promote economic activities to achieve more efficient functioning of the territory. That concept resembles, obviously, the French idea of *aménagement de territoire*. The Spanish Constitution only mentions the term once, alongside *urbanismo* (urbanism) and *vivenda* (housing) as responsibilities transferred to the regional communities. The Constitutional Court, by sentence 77/1984, stated that '*ordenación del territorio* and *urbanismo* have by goal the activity consisting of establishing the diverse uses to which land of physical territorial space can be devoted'. Responsibility for *ordenación del territorio* has been put entirely in the hands of the regional communities along with *urbanismo* and *vivenda*. This, in fact, had entitled the regional parliaments and governments to pass legislation and establish policy guidelines for the management of space.

Two sets of constraints exist limiting the freedom of movements of the regional communities. The first comprise the sectoral responsibilities for some basic resources, major infrastructure, economic and environmental policies, which remain in the hands of the national authorities, according to Article 149 of the Constitution. Co-ordination with these policies is essential for efficient structuring of the territory. Secondly, the existing *Ley sobre Régimen del Suelo y Ordenación Urbana* (herafter *Ley del Suelo*), formally published in April 1976, remains in force, even though those ideas, concepts and even full articles which are opposed to the constitutional principles subsequently approved in 1978 have for obvious reasons been superseded.

This law is the legal document of the highest rank currently valid throughout Spain dealing with detailed land use allocation: hence its importance for the purposes of this text. Several remarks should, nevertheless, be made in order to assess its actual validity. First, the *Ley del Suelo* was passed before the approval of the existing Constitution. There is a marked difference in their norms: the *Ley del Suelo* reflects a clear top-down, highly centralized government structure, whereas the

Constitution is the expression of a bottom-up, quasi-federal compromise between autonomous regions and a national state in which the regions have a much greater say, at least in planning-related matters.

Second, as a result of the allocation of responsibilities to national authorities referred to above, the regions have ignored the plans at strategic level (national and regional) envisaged in the *Ley del Suelo* and have drafted new plans at this level applicable to their individual territories. Those plans, in different states of definition within the 17 autonomous communities, provide policy guidelines for land uses, environmental protection and infrastructure development and, to a certain extent, may influence more detailed forms of planning.

Third, plans at municipal level (the important ones in terms of detailed specification of land uses and definition of actual development conditions) have stayed, with minor alterations, as they were originally conceived in the *Ley del Suelo*. Therefore, the analysis of their structure, as described in this basic law plus the regulations developing it, is still valid. That analysis is discussed in the second part of this chapter.

Some of the minor alterations mentioned above stem from the fact that final approval, exclusively with a purpose of overall co-ordination, now lies in the hands of the regional government, as opposed to the former situation where sanction from central government was required. This, obviously, means that some conformity between the municipal plans and an overall spatial (industrial) policy at regional level can be imposed upon the municipal government. Nevertheless, many municipal plans were drawn up and approved during the transitional period prior to the actual assumption of responsibilities by the regions, hence bypassing the theoretical overall co-ordination. Some municipal plans have had to be modified subsequently as a result.

Finally, the biggest metropolises in the country, with ample resources of technology and manpower, may have their own spatial or industrial policies which conflict with that of the respective regional government. This may even have political undertones if, as is the case in Madrid, political parties in the respective governments are of opposed sides of the spectrum.

In addition to the *Ley del Suelo* and its sister laws in the autonomous communities, there are several other sets of dual laws which affect spatial policy in general and industrial spatial policy in particular. Apart from laws relating to infrastructure (roads, land transport, airports) and laws dealing with natural resources (water, air pollution, natural areas), most of which stem both from national and regional parliaments, it is important to mention the *Ley de Bases del Régimen Local* which, among other things, deals with responsibilities of municipal authorities, one of the most important ones being planning at municipal level. The *Ley de Incentivos Regionales* sets up the overall framework, according to EC regulations, of industrial assistance in Spain. The *Real Decreto Legislativo* 1302/1986 is the transposition to the Spanish Code of Laws of EC Directive 85/377/CEE requiring environmental assessment of projects. There are, obviously, several other norms which one would like to consider when dealing with spatial policy in Spain. A fairly detailed list of the most relevant ones, as well as their spatial implications, may be found in Calderón (1988) along with additional references for further reading.

Summarizing this initial set of ideas, we can already establish some conclusions:

1. The administrative organization of spatial planning in Spain is a result of the prevailing constitutional framework, whereby three different levels of government co-exist: national, regional and local (municipal) with separate responsibilities of legislative and executive character. The Constitution splits up the sectors which fall within the responsibilities of national and regional authorities. *Ordenación del territorio* falls entirely in the hands of regional governments although other sectors (main infrastructure, basic resources, as well as economic and environmental policies largely able to influence spatial structure of the territory) lie under the aegis of the national authorities.
2. The legal document of the highest rank where physical planning is concerned is the *Ley del Suelo*, passed in 1976, prior to the approval of the Constitution. This law, if clashing in spirit with the prevailing bottom-up approach in policy-making, is still valid, and is widely used to draw up municipal plans, which control detailed land-use specifications. Most of the regional parliaments have passed their own equivalents to the *Ley del Suelo* and have produced, in accordance with their provisions, overall strategic plans or policy guidelines which may influence future municipal plans.
3. Certain other sectors affecting the spatial structure of the territory, upon which the national and regional governments do share responsibilities, have been the object of laws passed by both levels of Parliaments, often generating conflicts which are sorted out at the Constitutional Court.
4. The type of plan actually regulating development is drawn up at municipal level, in accordance with provisions set up in the *Ley del Suelo*. These plans are the exclusive responsibility of the local

authorities with an ultimate approval, for coordination purposes, from the regional governments. Certain conflicts have arisen as some of the most prominent local governments have spatial policies which may not coincide with those of the regional ones, mostly where each administration professes different political views.

4.1.2 Development plans and the control of development

This section aims to analyse municipal plans in detail. These are the basic tools of planning which allow subsequent development and set up the conditions whereby development may take place. The typology and contents of these plans will be presented in terms of the possibilities they open for industrial uses. The degree of conformity to specifications of those plans in order to obtain authorization for development, as well as the discretion, on the part of the municipal planning authorities to decide whether or not to authorize development which does not fall exactly within the plan specifications, will likewise be assessed. Procedural details for obtaining permits, plus development control, rights and liabilities of the developer and an assessment of the rigidity of the system, will complete this part.

Municipal plans

As has already been stated, the municipal plan is the planning document which sets up the conditions for development. Its drawing and approval are the responsibility of the local council (*ayuntamiento*) with a final sanction from the regional authorities. The main features of the municipal plans are described in the *Ley del Suelo* and *Reglamento de Planeamiento*, with additional details regarding plan management and control of development in the *Reglamentos de Gestión* and *Disciplina Urbanistica*. The law refers to the municipal plans as strategic in character as they establish the basic structure of the municipal territory. This statement should be clarified, since specifications of the plans vary between the different categories of land which the plan itself establishes, as we will explain further below. There are three different types of municipal plan: *plan general municipal de ordenación* (PGMO); *normas subsidiarias de planeamiento municipal* (NSPM) and *proyectos de delimitación de suelo urbano* (PDSU). Although the main goal of all of them is to establish the prevailing conditions for development, by zoning, in the whole of the municipal territory, there are important differences between them. The most obvious one is their degree of complexity, maximum for the

PGMO and minimum for PDSU. Sensibly enough, the election of one or the other depends on the relative importance of the municipality concerned: municipalities with big cities or provincial capitals should go for PGMO whereas those with no development should opt for PDSU. In between lie the rest of the municipalities with varying degrees of current or expected developments, which should opt for either of the two alternative forms of NSPM, to be described below.

For the top category of municipal plan, the PGMO, the broad categories of land use the plan may establish are as follows.

1. *Suelo urbano* (urban land): Land classed within this category is directly usable for development, as it is consolidated or has vehicle access, infrastructure for water supply and disposal, energy supply, etc. The plan itself will indicate, for this broad category of land use, the detailed uses (*usos pormenorizados*) and levels of intensity of those uses in terms of buildings per hectare, building typology, permitted uses for buildings, etc, which may be authorized. No further planning is required for this category of land unless otherwise stated in the plan. In that case, the *plan especial de reforma interior* (special plan for interior reshaping) for one particular sector, will add on to the broad land uses defined in the PGMO, the detailed ones, with the same level of detail as those in the PGMO.
2. *Suelo urbanizable* (land suitable for devlopment): This land requires further planning and infrastructure provision before it can take buildings. The PGMO may establish two further categories of *suelo urbanizable* by attaching time-tags to it. *Suelo urbanizable programado* sets up two four-year periods during which this land must be converted into *suelo urbano* and thus put in the market for development. For the *suelo urbanizable no programado*, there is no commitment from the local administration as to when the land will be made ready for development. It is, therefore, a reserve of land the *ayuntamiento* holds to offset or foster demand as it wishes. To become *suelo urbanizable programado*, a programme is required in which all administrations involved commit themselves to the provision of the lacking infrastructure, and the two four-year periods will be established. No programming is required for either NSPM or PDSU.

   In terms of additional planning, *suelo urbanizable* requires a *plan parcial* (partial plan) which, apart from establishing the physical structure of the sector concerned, will mark the *usos pormenorizados* and their intensity, taking into account the broad ones the PGMO will have to have marked for that sector.

3. *Suelo no urbanizable* (land unsuitable for development): The classification of an area as *suelo no urbanizable* means the *ayuntamiento* wants to preserve it from development, either to devote it to agricultural uses or simply hold it by virtue of its environmental quality. The plan will indicate the range of agricultural activities and associated buildings permitted or the level of protection allocated to the area. In principle, the only possibility of developing *suelo no urbanizable* is by drawing the whole plan anew, with new goals and objectives, or modifying parts of it, but in either case subject to all the procedural steps for approval the law requires for a new plan.

As for the remaining types of municipal plan mentioned above, the features of each of the categories are exactly the same. The NSPM cannot create *suelo urbanizable programado*; in other words, there is no time commitment from the *ayuntamiento* in terms of development and just the three main categories of land may be established. There is, notwithstanding, an abridged version of NSPM, for small municipalities with limited prospects of development, in which only *suelo urbano* and *seulo no urbanizable* are established. As before, the *suelo urbano* may be further planned by way of a *plan especial de reforma interior*.

It will be obvious that the borderline between this abridged NSPM and what has been defined above as PDSU is very thin. PDSU is the minimum type of municipal plan required by the *Ley del Suelo*. It simply reflects the current situation of the municipality as no perspectives of development are envisaged. The PDSU will indicate what part of the territory is considered *suelo urbano*, and the building conditions in it; no further planning can be drawn here. The rest will be considered *suelo no urbanizable* and thus preserved from development.

We have now, examined the different types of municipal plan envisaged in the *Ley del Suelo*. At present, most of the 8,000-plus municipalities in Spain have one or other of these plans, and some of them have already undertaken revision of them in view of changes in the circumstances which were initially established. Once a sector of land is classed as *suelo urbano*, either directly by the *plan municipal* or after a subsequent piece of detailed planning already described, development conditions are rigidly set up for that sector. In the rest of this section we shall deal with building licences, rights and liabilities of land owners and the rigidity of the system.

According to Article 1 of the *Reglamento de Disciplina Urbanística*, authorization must be sought from the municipal authorities, among other things, to:

1. construct new buildings;
2. enlarge existing ones;
3. modify the structure of existing buildings;
4. modify the layout of a building externally or internally;
5. carry out temporary works to be demolished later;
6. change the use of buildings;

as well as any other activity listed in the municipal plan, or any other plan elaborating it. Article 3 of the same *Reglamento* states that licences will be granted 'in accordance to provisions in *Ley del Suelo*, municipal plans . . .' Both authorization and refusal to develop must be justified by way of conformity or otherwise of the application with those provisions.

Additional authorizations from the competent authorities may be required if infrastructures or natural resources whose management escapes the municipal responsibility are going to be affected. In that respect, the share of responsibilities established by the Constitution, as well as provisions in the *Ley de Bases del Régimen Local*, should be taken into consideration. A similar comment may be made with regard to development proposals which entail a significant impact on the environment, according to the annexes of the *Real Decreto Legislativo* 1302/1986.

We might, therefore, conclude that the Spanish system of development control is rather rigid. This is true; but there are ways of introducing flexibility into the process. A first, obvious, point relates rigidity to the level of detail with which *usos pormenorizados* are defined in *suelo urbano* or, by way of a *plan parcial*, in *suelo urbanizable*. There are no precise instructions in either *Ley del Suelo* or *Reglamento de Planeamiento* about basic typologies of *usos pormenorizados*, how detailed the definition of those uses should be, or incompatibilities between them. Planning practice has established some terms which have become more or less customary in the jargon. For industrial uses, detailed uses are specified by conditions of density, volume, energy consumption, etc. Specific limitations in terms of incompatibilities between industrial and residential uses can also be introduced.

A *plan parcial*, in a sector of *suelo urbanizable* broadly classed as of industrial use, can be initiated by the prospective developer, be they private or public. One or the other may willingly be relatively lax in their definition of *usos pormenorizados* either to accommodate one particular type of industry or to attract industry in general without much hindrance in terms of restrictions of use. This is the case of *polígonos industriales*, sited in specific locations in many municipalities, divided in plots and with basic infrastructures provided by the municipal authorities or even the regional ones in order to foster industrial development in that particular location. Sensibly enough, changes in *usos pormenorizados* are more cumbersome in *suelo urbano* where densities are higher and uses are more mixed. Incompatibilities between uses become a crucial issue as do all environmental considerations.

A change in *usos globales* would require the complete revision of the municipal plan, or a specific modification of the broad land uses. In either case, the administrative procedure is quite similar to that of the initial formulation of the plan. One exception to that rule is established in Arts 85 and 86 of *Ley del Suelo*, with regard to limits to the development of *Suelo urbanizable no programado* and *suelo no urbanizable* in a PGMO, an exception which could also be extended to *suelo urbanizable* and *suelo no urbanizable* of the remaining types of municipal plan examined above. This exception allows uses other than those directly related to agriculture, including industrial activities, on condition of a previous consideration of those activities as of 'social interest'. This is a fairly straightforward procedure the *ayuntamiento* itself may undertake, which has led to certain abuse in municipalities where job creation has been considered an issue of social interest.

As an initial set of conclusions to this section several issues may be brought forward.

1. Authorization for development is based strictly on the provisions of municipal plans and is granted by municipal authorities. Complementary permits may have to be sought from other administrations responsible for infrastructure or resource management.
2. The system is as rigid as the detailed provisions in the plan.
3. The possibilities of changes in detailed uses (*usos pormenorizados*) are related to the lack of precision in the definition of permitted uses.
4. The possibilities of changes in broad uses (*usos globales*) require a total revision of the plan or piecemeal modification, in both cases undergoing all steps in the standard process of plain formulation.
5. A loophole in *Ley del Suelo* has enabled some municipalities to authorize industrial development in zones classed as *suelo urbanizable* or *suelo no urbanizable*, by declaring development proposals as of 'social interest'.

Rights and liabilities of the developer

The rights and liabilities of the developer are the subject of Title II of *Reglamento de Gestión Urbanística*. The main right of the developer (the developer is for the time being equated with the owner of the land) is, not surprisingly, to develop; or in other words, to carry out any of the operations listed in Art. 1 of the *Reglamento de Disciplina Urbanística* mentioned above. *Ley del Suelo* clearly specifies that this right should be exerted within the terms of this law and the regulations developing it, as well as within the limitations included in the municipal plan which affects the land subject to the prospective development. A second right enjoyed by developers within the context of the plan refers to a fair share in terms of benefits and liabilities as the result of the plan, as it is obvious that not all zones included in the plan end up with the same quantitative rights to develop as a result. To equate the developer and landowner excludes, obviously, other forms of tenure or contractual relationship between parties to be dealt with in section 4.2. Additional rights and liabilities between owner and developer, if they are not one person, will be regulated in their contract and presided over by specifications of *Decreto Ley de Arrendamientos* which we will examine further in section 2, as it affects supply and demand of industrial property, length of contracts and possibilities of changes of use.

The liabilities imposed by the *Ley del Suelo* and the municipal plans on the developers are listed in Arts 83 and 84 of the Law and can be summarized as follows.

1. They must give away to the *ayuntamiento*, free of charge, all land needed for public rights-of-way, gardens, schools, etc. in *suelo urbano*. In *suelo urbanizable*, land for other public services such as sports grounds, cultural centres, etc, which do not already exist at the time of the development must also be handed to the *ayuntamiento*.
2. Owners of *suelo urbanizable programado* must give, in addition, 10% of their building rights, once the plan, by way of a standard calculation, has averaged that building right for all owners of land in each development sector. Those building rights the *ayuntamiento* obtains free of charge are meant to endow the *ayuntamiento* with a patrimony of land to be used to provide social amenities or available land for development as it wishes.
3. The developers must build all the necessary basic infrastructures, as listed in Art. 122 of *Ley del Suelo*, before the actual land development

starts, and meet all the costs incurred, including indemnities due and the cost of additional planning and project work.

4. The developers must develop the land within the time limits marked in the plan; if this clause is not met, the licence expires without the developer being entitled to any kind of compensation.

All the above conditions are liabilities of a general character which stem from the *Ley del Suelo*. Additional ones may be imposed by the plan aimed, usually, at the protection and enhancement of items of interest in an urban environment, but which may also imply modifications of the appearance of an industrial building, its architectural features or state of conservation.

4.1.3 Financial issues

Title I of *Reglamento de Gestión* refers to the main actors in the actual execution of the provisions of the municipal plans. Remember that some articles in *Ley del Suelo* and its subsequent *Reglamentos* have been altered as a consequence of the approval of the Spanish Constitution. Allowing for such alterations, and the fact that the regional governments have taken on their share of responsibilities in the development process, a simple classification of the main actors which intervene in it would highlight two main groups:

1. central, regional and local administrations;
2. private initiative.

As a general principle, the three administrations can act individually or jointly in accordance with *Ley de Bases del Régimen Local* and their own governing norms; alternatively, they may form special entities to carry out development.

As mentioned previously, the municipal (local) administration is the one with a more specific responsibility in the execution of the municipal plan. The *ayuntamiento* may assume that task through its ordinary specialized staff or, in big municipalities, by creating a decentralized unit called *gerencia de urbanismo* (see *Reglamento de Gestión* section 4). This unit may carry out, among other things, the infrastructure works required to develop an industrial site.

Municipalities can, likewise, create special bodies outside their own administration aimed at land development. The most common ones are *consorcios urbanísticos* (see *Reglamento de Gestión* section 3) in which different administrations and even private individuals may link forces to reach common objectives in which they all have interests and responsibilities. Those objectives range from studies, to site development, detailed plans, assistance to the municipal administration to carry out provisions of the plans, to build urban infrastructures, etc.

Finally, the municipal administration may form *sociedades anónimas*, ruled by private law, along with other public administrations or private institutions (see *Reglamento de Gestión* section 5). These *sociedades anónimas*, in which the administration must hold the majority of the shares, can undertake infrastructure works, manage and promote activities in industrial zones, etc.

When required, several municipalities may join forces to undertake certain tasks, among them those related to spatial planning and site development. The municipalities concerned may enter these groups voluntarily (*manconumidades*) or may be compelled to do so by the regional or provincial administrations (*agrupaciones forzosas*). The rules which govern these entities are normally agreed upon by the parties involved, which obviously have to relinquish a part of their individual responsibility for the benefit of the group. As stated above, the municipal administration is the main public actor in development. Without encroaching upon its responsibilities, the other two main administrations, central and regional, may carry out their duties in much the same way as the municipal one, by themselves or creating analogous structures, except *mancomunidades* and *agrupaciones*. Elsewhere in this chapter we shall examine some examples of the role of these two administrations in promoting development.

Article 2 of *Reglamento de Gestión* clearly states that private initiative is the most important actor in development; and the administration should just concern itself with overall control, protection of the general welfare and to see to those issues which escape the individual interest. The private developer may undertake detailed planning, in accordance with provisions in the municipal plan, and may act in site development by forming owners' associations with or without the presence of the municipal administration, in accordance with provisions in *Reglamento de Gestión* (see section 6, and Title 5).

Once the main actors in development have been named, a further issue of interest concerns the financial provisions for development. We have already examined how the *Ley del Suelo* and *Reglamento de Gestión* compel the private developer to meet the costs necessary to upgrade the quality of a sector suitable for development when certain infrastructures are lacking in it. At the same time, the private developer is responsible for all costs associated with the development itself.

The role of the administrations can be summarized by saying it is meant to encourage development of the provisions in the plan, either by carrying out their own responsibilities or by taking part, under either of the entities mentioned above, in the actual development process. It is the second of these options that holds more interest in relation to the objectives of this book. We have mentioned three main types of entity through which the public administrations at central, regional and municipal level may intervene in the development process. We have also outlined their characteristics and the duties each of them may perform to foster development. Out of those three, *gerencias*, being very common in large municipalities, make part of the municipal organization and, therefore, may be counted within the first option above.

As for the *consorcios urbanísticos*, we have seen that they may, in carrying out development,

1. undertake promotional works for particular industrial sites;
2. carry out infrastructure works;
3. create or manage ancillary services for industrial activity.

A typical example of a multi-sided *consorcio* now in progress has been formed between *Comunidad Autónoma de Madrid* (regional administration), *Ayuntamiento de Las Rozas* (municipal administration outside Madrid), and a private consulting firm, for the promotion of a science park in that particular municipality.

By far the most popular type of entity to foster industrial development is the *sociedad anónima*, ruled by private law, which grants a lot more flexibility than the administrative law that rules public activities. Nevertheless, the municipal administrations making part of a *sociedad anónima* must hold majority voting power, so as to preserve the public interest. Options open to this type of *sociedades anónimas* are defined in similar terms as the ones described for *gerencias* and *consorcios* in Art. 21 of *Reglamento de Gestión*. But the real possibilities stem from the very character of the *sociedades anónimas* which may create their own objectives within the broad framework established by the *Reglamento*.

IMADE, *Instituto Madrileño de Desarrollo* (Institute for Development of Madrid) is a public entity ascribed to the regional government, its main formulated goal being 'to promote the development of the productive system'. It meets this goal by forming *sociedades* with other entities, public and private alike, which supply the capital. The objectives of these *sociedades* are quite diverse and their character may be financial, advisory, or technical, but in all cases there is a financial contribution from the regional budget to the joint venture, be it through the regional government or by the regional savings banks controlled by the government.

Another good example of *sociedades anónimas* with aims of regional development are the so called SODI (*Sociedades de Desarrollo Industrial*) made up by *Institute Nacional de Industria* (National Institute for Industry), which holds the majority of the capital, the rest being topped up by savings banks, regional governments, etc. Their aims include financial participation in private groups for industrial development. Similar *sociedades* have been created within the autonomous regions.

The financial contribution of any administration to industrial development is subject to restrictions from the EC. There is no need to repeat those restrictions as they are common to all member states. In order to offset some of those monetary controls and as a way of diversifying their means of intervention, public administrations are using alternative schemes to foster industrial development, which amount to financial outlays or savings in the development costs to be met by the industry. Some of these alternative schemes are:

1. one-stop permit process, to ease the process of making applications for development;
2. business assistance programmes, to inform prospective developers about all types of industrial aids in one particular location or to offer technical assistance to firms in distress;
3. creation of amenities (including low-cost rental rates, free land, shared services, office and parking space) to attract small industries;
4. low-cost financing;
5. lowering/waiving of creation costs and tax rebates.

All these are geared to attract private developers to particular locations.

4.2 INDUSTRIAL PROPERTY

The second section of this chapter considers the market process for industrial property in Spain. The aim of this section is to analyse mechanisms for development of industrial land and access to industrial property. Occupation and rights of occupiers will be then discussed. Thirdly, market prices and rents, followed by public intervention mechanisms, will be discussed. Market operation is then described in terms of efficiency and adjustment. Finally, market responsiveness and the current situation of industrial property market in Spain are referred to.

An initial snag is the lack of reliable, accessible information about the industrial property market in Spain. This may be due to the youth of the activity itself or, even, the accession to the market of investors seeking profits which they do not want to share with others. The result is a void in statistical and factual information which may render difficult a cross-comparison between the Spanish and other European markets.

4.2.1 Industrial land development

The process of industrial land development, and thus the creation of new supply, has already been described in section 4.1 from the planning viewpoint. Because of the outburst of industrial land allocation in the last decade, reference is made primarily to new developments, although occupation of existing industrial properties will also be mentioned.

The standard process is for industrial developers to approach the municipal council seeking *suelo urbanizable* broadly classed as of industrial uses. Industrial development attracts the interest of local politicians as a source of employment. That may, in turn, lead towards changes in global uses, according to the procedure described above in order to accommodate supply to an existing demand or, in many cases, may stimulate that demand regardless of the plan provisions. These reforms have been a common feature in the location of large multinational firms in Spain. They have also been the solution when provision of *suelo industrial* has fallen short of actual demand.

Ayuntamientos expecting or trying to encourage industrial development provide in their local plans, *suelo industrial* in large land units known as *polígonos industriales*. Large numbers have been created in the 1980s. The general trend is for provision of self-contained *polígonos industriales* away from the urban areas, since local plans aim to remove all industrial sites from the inner city. However, many of these municipal councils are relatively limited in their managerial capacity and they find themselves unable to carry out development by themselves; hence the important role of specialized organizations involved in industrial development. In fact, the term *polígono industrial* derives from the name applied to those industrial promotion nuclei promoted in the 1960s and 1970s within the National Economic Development Plans.

As we have hinted above, some of these industrial public promotion agencies have specific responsibilities for industrial land development at national (such as SEPES, an autonomous agency of the national Ministry of Public Works; see below) or regional levels (for example in the Basque Country, Catalonia, Andalusia and Galicia), depending on the degree of devolution achieved by each regional community. In addition, other public organization (such as the Departments of Industry at regional level) are also involved in the process through the provision of finance, expertise, promotion and information. There are also private estate agents involved in industrial land development, but not many display a national dimension and some local ones are short-lived as they are established for the development of a specific industrial site only.

Both public organizations and private developers generally arrive at an agreement with the municipal council for the development of industrial units. The terms of these agreements vary depending on the size of the project, employment creation, land classification and municipal land management capability, sometimes involving the adjustment of the local plan to the developer's needs.

Industrial estate agents as well as public agencies tend to acquire the land and develop the industrial site, or *polígono industrial*, with the basic infrastructure and services in accordance with provisions in *Ley del Suelo*. The development is then divided into individual plots and sold to private firms who build their own premises in accordance with the detailed regulations contained in the local plan for the area. These may determine, for instance, the type of industry which is acceptable, the physical form of the development, etc.

The *Registro de la Propiedad* is the official registry for property in the municipality. A file is kept for every individual estate (land and buildings) including its description, name of owner, price of acquisition, occupation, taxation liabilities as well as all the sales, changes of ownership, and subdivision or aggregation of the plot.

4.2.2 Occupation and rights of occupation

Industrial firms in Spain have traditionally acquired land and built facilities to suit their own needs, owner-occupation being a general feature and a characteristic of the Spanish traditional entrepreneur class. However, during recent decades there has been a rapid increase in renting as opposed to purchase, although this still only supplies a small proportion of the market and is more common in large urban areas. This new trend has been encouraged by the recent boom in market prices and the introduction, as foreign investors stepped into the market, of new systems of tenure. The boom prompted the appearance of a new actor, the private industrial estate agent, operating in parallel with the traditional public agencies for land development, but who, unlike the latter, rents industrial units. Some other public organizations (such as municipal councils and regional governments) rent and lease industrial facilities under special conditions, as we have seen above, as part of their industrial promotion strategy, although this is still exceptional.

Letting contracts were regulated by *Ley de Arrendamientos Urbanos*. This law protected the occupier's security of tenure, hampering the owner's ability to evict him or alter the terms of their contract. Recent amendments to this law, introduced in the *Real decreto-Ley sobre Medidas de Politica Economica* of April 1985, have removed tenants' rights to an automatic extension of the contract. This has forced a higher mobility on non-owner-occupiers and the development of new forms of letting contracts. Finally, unlimited rights of purchase and sale are generally guaranteed to all sites and premises. There are situations where the different administrations have the right of expropriating due to public interest which has to be established on 'fair' grounds. The estate is then acquired by the administration at its official price as stated in the *Catastre* (see below) plus a compensation allowance according to the activity in the estate prior to acquisition.

In relation to rights of occupation and use, possible changes in the use of industrial facilities and/or in its physical condition (new buildings, refurbishment, demolition, extension, etc) are regulated by the local plan in accordance with Art. 1 of *Reglamento de Disciplina Urbanística* (see above). In practical terms, the enforcement of local plans is stronger in new developments by virtue of the licence-issuing procedure, whereas minor physical changes and extensions are difficult to control; when changes are illegal, a fine can be applied, but demolition (as a strict interpretation of plan regulations demands) is rarely carried out.

As mentioned before, the level of detail in regulations varies greatly among plans. They generally refer to parameters such as maximum building volume, and height and/or distance from perimeter of site, but can also include maximum number of windows and even colour. Permitted and not permitted activities (uses) are generally, with some exceptions, described in general terms (such as non-polluting industrial activities, or industry). Thus a detailed plan regulation can constrict occupation and use to varying degrees. Moreover, industrial facilities must comply with national and regional legislation as well as with the special by-laws issued by the local council.

4.2.3 Property price and rent

The *Catastre* is the general registry for real estate property which establishes its price and condition and serves as a reference price for taxation purposes (VAT, property tax, and local rating for purchase) as well as for public acquisition.

Purchase price is not regulated, except when public acquisition or confiscation take place, but it is indirectly affected by:

1. estate taxation;
2. local plan provisions for the area in terms of infrastructure or services.

Often there is a duality in the property price which combines the official price, stated in the Estate Property Registry, and the actual purchase price. The former determines tax rates (local rates and property taxes), the latter being generally higher. This duality, although illegal, has been a common feature in the Spanish property market during the last two decades. The enforcement of general taxation control by the Treasury (since the mid–1980s) has forced 'black' money (i.e. non-declared incomes) to come onto the estate property market. Nevertheless, enforcement of tax control on real estate marketing activity in the late 1980s, and the current updating of *Catastre* will continue to make official and actual prices come closer together. Currently, (January 1992) the *Catastre* valuation process is being delayed by court decisions about the validity of its procedure.

Local plan provisions for the area (and thus permitted uses and physical standards for land and buildings) affect property values by defining development potential. This has become decisive since town planning became a common practice and the enforcement of local plans and development control followed as a result of the political and administrative changes including greatly increased municipal autonomy, occurring after 1976. The recent boom (1983–90) in the estate property market has, in some cases, also affected values of land classified as *suelo no urbanizable*, due to the underlying development potential which may be brought about as we have described above.

In relation to rents, the *Ley de Arrendamientos Urbanos* of 1964 established that rent contract renewals could only increase the rent up to a statutory maximum level which resulted from applying the official index for life costs increase (known as IPC, issued annually by the National Institute for Statistics) to the previous rent. Rents for new contracts are not regulated. This protective legislation was seen by the Socialist government as a brake on economic development, and thus the 1985 amendment was passed removing security of tenure. This has produced a dual rent market: while contracts signed before May 1985 enjoy regulated prices and security of tenure, lettings after that date have seen their rents increase dramatically as property prices rose during the 1980s, forcing a higher mobility on occupiers. This duality has had a dramatic affect on housing, with a minor impact on the industrial renting sector, due to the lack of a renting tradition among Spanish firms.

4.2.4 Public intervention

The main role of local planning in industrial land development and subsequent changes in the use of industrial properties has already been described. Financial assistance and incentives for capital investment in industrial development projects (including industrial land development and acquisition of industrial facilities) in most regions of the country, according to provisions in *Ley de Incentivos Regionales* and ERDF regulations. Public finance may be obtained for industrial estate development through different schemes ranging from the preferential industries promotion scheme to special credit from public banks. In addition, the regional communities have developed their own instruments for industrial promotion and provide financial help as well as information and advice on the industrial estate market.

Finally, we have already mentioned the main role of public bodies for industrial land development such as SEPES (*Sociedad Estatal para Promoción y Equipamiento de Suelo*) and its regional parallels in the regional communities of Galicia, Andalusia, Basque Country and Catalonia. SEPES is an autonomous public agency associated with the Ministry of Public Works which acquires and develops industrial land (recently it has extended its role to tertiary facilities), providing infrastructure and services, in order to sell it into the open market. In all cases, it comes to an agreement with the responsible local council and can also undertake joint initiatives with private developers along the lines described above. SEPES inherited a large amount of publicly owned land from the previous land development agency at the former Ministry of Housing, thus reaching a prominent position in industrial land development once the crisis was overcome early in the 1980s.

Local authorities can apply for SEPES aid to finance their industrial land development programmes. Its actions are selective in nature since it combines its public character, and thus regional policy goals, with its autonomous status and so its need to reach an economically positive balance in its annual operations. Although most of its initial work was based on the development of vacant public land within the old *polígonos industriales*, it has now focused on the main growing areas such as the Mediterranean ribbon and the Ebro axis. It is also operating in areas such as Castille and Galicia where private initiative finds it difficult to develop industrial land.

For the sake of completeness, reference should also be made to the nationalized industries, *empresas nacionales*, as industrial estate occupiers (consumers). The group of large national firms are generally industries demanding specific industrial plants, but some also occupy standard industrial facilities (storage, high-tech, retailing). Although the behaviour of *empresas nacionales* is not always adjusted to strict market operation, and their facilities are not always easy to include in the pattern of industrial facilities considered here, their interest lies in their capacity to attract ancillary industrial development by the sheer size or type of their operations. A typical example would be the wholesale food and vegetable markets located in large areas outside many metropolitan areas of Spain.

4.2.5 Market operation

The characteristics of the supply side of the industrial estate market have gone through a change associated with both the evolution of estate market values and the growing demand from small and medium-sized firms. Thus the late 1970s and 1980s witnessed a decline in the average size of industrial facilities which has, in recent years, increased slightly due to the development of retailing facilities. In the same period modern *polígonos industriales* were developed, smaller in size than the large traditional ones, generally located in the outskirts of towns, and close to main roads. The acquisition of land by developers for building their own facilities is still dominant over letting. In relation to facilities for sale ready to occupy, the supply of standard modules of 250 m^2 which can be acquired jointly (500 m^2, 750 m^2) seems to be the trend in the large urban areas.

On the other side of the market, demands from new users respond to the types of new industries emerging (small and medium-sized manufacturing, food and drinks, agricultural processes; also storage, retailing, logistic industries, etc). Thus users require facilities ranging from small workshops to medium-sized units, including basic services and attractive design. They pursue locations close to transport links, but this is not so widespread as it used to be, and depends on the type of industry and its transport requirements.

Generally, industrial property actual values adjust freely to market except for the intervention of local planning policies and public land-developing agencies. Local planning policies, although generally in pursuit of market adjustment, can reduce flexibility of adjustment due to their provisions in terms of land available for industrial use, and their associated bureaucracy. In any case, they play a main role in the more dynamic areas such as Madrid and Barcelona. Public agencies for land development extend the supply side by the very nature of their operations, and their intervention can often break 'bottle-neck' situations and stimulate the market.

The recent appearances of estate investors, and thus of speculation among industrial land and facilities (a side-effect of the estate price boom of the last years), seems negatively to affect the market operation. Currently, the dramatic increase in industrial estate prices, whose high growth rate had slowed down slightly during 1990, has affected negatively the already growing demand which starts to find it difficult to afford the new prices.

Although market information is collected in the *Registro de la Propiedad*, problems with updating and accuracy of information have made the registry inadequate for this role. Generally, consumers get information from entrepreneurs, associations and public promotion agencies. These agencies, which are part of the regional communities administration, have recently included information services on industrial estates' availability among their promotion schemes. For example, IMADE (Madrid Institute for Development) has updated its information every three months since January 1989. In addition, the new industrial estate agents advertise their supply in specialized magazines as well as in regional television networks. Nevertheless, availability of reliable information is still a problem.

On the letting market, the removal of security of tenure and of statutory duration of contracts, coupled with the increase in property prices, has resulted in a greater mobility of rents both upwards and downwards, and an upward adjustment of them to match the steady increase in market prices.

Industrial unit specifications, basically size and ancillary facilities, provided by the supply side of the market have progressively adjusted themselves to the new characteristics of demand as the market has become more dynamic during the last decade. Thus the average unit size had decreased while the quality of building materials and the aesthetic appearance of the main ancillary buildings has improved. The general modernization of the small and medium-sized enterprise (SME) sector, the main consumer in unit terms, has encouraged developers to improve quality, while the high benefits to be gained have furthered the provision of high-quality facilities. However, the steep increase in market prices and rents is creating changes as prospective occupiers find it difficult to afford prices for the quality they require, seeing themselves forced to opt for cheaper facilities and lower standards in terms of size, services or location.

4.2.6 Market responsiveness: recent trends

The Spanish estate property market has undergone a dramatic boom during the late 1980s as a result of several factors. Economic recovery of the country, attraction of foreign investment, restrictions on available land imposed by conservative town-planning strategies, the enforcement of taxation control measures and the removal of automatic extension of tenure by new legislation have been the main causes for the spectacular rise in estate property prices during the 1980s.

This has been more dramatic in the large and medium-sized urban areas (Madrid, Barcelona, Seville, Valencia) as well as in coastal resorts, although it can be considered a general phenomenon throughout the country. The sudden price increase has gone mainly to housing and offices, but lately it has extended into industrial property prices. Foreign estate investors have been attracted to the domestic market, and although their main interest has been in tertiary facilities and, on a smaller scale, in housing, they have recently entered the industrial estate market.

The Spanish structure of industrial firms has traditionally composed a large sector of small and medium-sized firms and a few large companies. The former have recently developed their production systems as a result of the economic adjustment after the crisis, so they have come into the industrial property market looking for more suitable facilities than their traditional ones. These firms have a strong local character; their mobility is limited to a specific urban area and, in some regions, to wider geographical areas (*comarcas*). They are mainly in the manufacturing (textile, shoes), agriculture processing and food and drink sectors. Yet there is a small but active group of high-tech firms which had developed rapidly from traditional manufacturing sectors. The retailing sector has recently shown a quick and far-reaching expansion and, although they enjoy ample mobility at regional and national scale, their facilities seek locations within selected areas such as main urban areas. Only large firms have mobility above the regional scale and they often involve specific facilities (production plants).

The structure of the Spanish regions and particularly the urban settlement pattern provides for a widely fragmented and diverse industrial property market. Some areas are highly populated, industrialized, with good internal communications. These enjoy an active industrial property market due to the diversity of sectors involved and the flexibility of location. The Madrid and Barcelona areas are regions with a traditional small and medium-sized firm basis and fall within this category. Traditional industrial areas such as Santander Bay, centre of Asturias and the Basque Country have large urban areas and are at present undergoing a re-industrialization process which has activated, albeit timidly, their traditional industrial property estate, bringing smaller units into the market serving especially the retailing, agricultural produce and food and drink sectors.

Other areas along the Mediterranean littoral (Valencia, Murcia, Andalusian Coast) and the Ebro river axis (Zaragoza, Navarra, Rioja, interior of Catalonia), the most dynamic areas in terms of growth rates, are developing industrial facilities in a more demand-responsive manner, according to the characteristics of the firms (food processing, manufacturing, agricultural processing and storage, catering for tourism), in many cases of a certain local character, which want to establish themselves in the area.

Finally, industrial indicators show smaller growth in the extensive central part of Spain (between Madrid and the coasts), with some industrial areas, usually around medium-sized towns, generally administrative capitals and/or main agricultural centres. They have medium and small, but active, industrial property markets (Valladolid, Aranda de Duero, in Castilla y Leon; Alcazar, Albacete, in Castilla La Mancha; Cordoba in interior Andalucia), drawing upon the potential of their own area with the benefits of location on major transport routes.

APPENDIX: GLOSSARY AND LEGAL INSTRUMENTS

Ordenación del territorio

The meaning of this concept has been the subject of debate amongst experts, its goals swinging from a pure economic activity aimed at fostering regional economic development to a more executive endeavour in which land uses and infrastructure provision are coordinated to gear population and the economic activities for a more efficient functioning of the territory. That concept resembles, obviously, the French idea of *aménagement de territoire*. The Spanish Constitution only mentions the term *Ordenación del Territorio* once, alongside *urbanismo* (urbanism) and *vivenda* (housing) as responsibilities transferred to the Regional Communities. The Constitutional Court, by sentence 77/1984, stated that '*Ordenación del territorio* and *urbanismo* have by goal the activity consisting in establishing the diverse uses to which land of physical territorial space can be devoted'.

Ley sobre Régimen del Suelo y Ordenación Urbana

Its text was published by *Real Decreto* 1346/1976. Its closest British equivalent is the Town and Country Planning Act. It is the legal document of the highest rank currently valid in Spain in the area of physical planning. Its articulate has been subsequently detailed by means of three *Reglamentos* (regulations) dealing, respectively, with types and contents of plans (*Planemiento*), management of plans (*Gestión Urbanística*) and development control (*Disciplina Urbanística*).

Plan general municipal de ordenación

General municipal plan of strategic character for the whole of a municipality which includes the capital city of a province or any other big or fast-growing city.

Normas subsidarias de planeamiento municipal

Subsidiary norms for municipal planning. As *plan general-municipal de ordenación* but for less important or slow-growing municipalities.

Proyecto de delimitación de suelo urbano

Project to mark urban land. As above, but for municipalities, with no perspectives of growth.

Ley de Bases del Régimen Local

Basic law on local government, passed 2 April 1985: the basic law which regulates the functioning of all administrations below the regional level, their responsibilities, organization etc.

Ley de Incentivos Regionales

Law on regional incentives, passed 23 December 1985, and the *Reglamento* to develop it, approved by Royal Decree RD 1535/1987 of 12 December. As a result of the Spanish accession to the EC, it establishes, in accordance with EC principles, the ceilings of aid which may be granted to industrial and infrastructure development throughout Spain.

Ayuntamiento

Municipal council, formed up by elected councillors

Polígonos industriales

Industrial sites, provided by either private or public initiative, with a structure approved by a *plan parcial*, divided into plots and with basic infrastructures built in.

Gerencia de urbanismo

Decentralized unit within the municipal administration dealing with land use planning and development.

Consorcios urbanísticos

Special units outside the administrations formed by one or more of those administrations plus, in some cases, private institutions, to carry out objectives of common interest.

Mancomunidades de municipios

Voluntary grouping of several municipalities to carry out specific tasks, among them planning and development.

Agrupaciones forzosas de municipios

As above when imposed upon the affected municipalities by a higher administration.

Registro de la propiedad

A registry of all property whose information supplements the *Cadastre*.

Ley de Arrendamientos Urbanos

Law on letting of urban properties, approved by Decree 4104/1964 of 24 December, and modified by *Real Decreto-Ley* 2/1985 of 30 April on economic policy measures.

Empresas nacionales

Large publicly owned industrial firms in several key sectors.

REFERENCES AND FURTHER READING

Calderón, E. J. (188) *Lecciones de Ordenación del Territorio*, Vols. 1 and 2 (Lecciones Nos 6 y 28). Servicio de Publicaciones, ETS Ingenieros de Caminos, Universidad Politécnica de Madrid.

García de Enterría, E. and Parejo Alfonso, L (1981) *Lecciones de Derecho Urbanístico*, Editorial Cívitas, Madrid.

5

FRANCE

Alain Motte

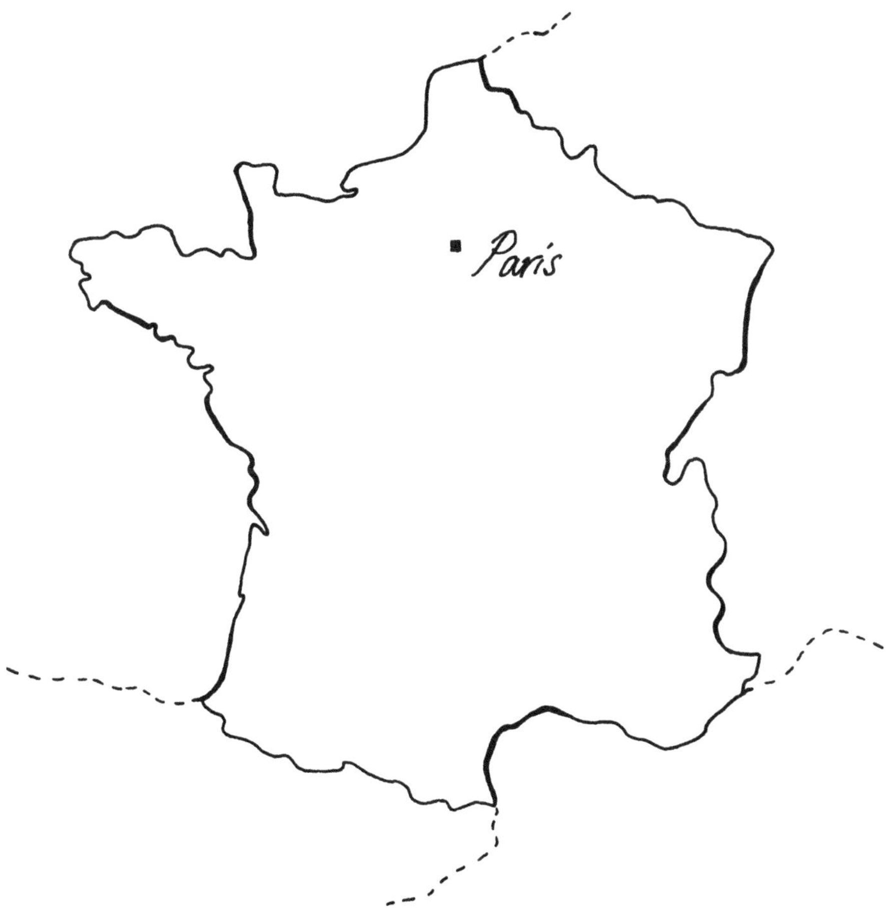

This chapter aims to give a general presentation of the nature of the industrial property market in France and to show the role of town planning in its structure and operations. Two main ideas underlie this presentation: first, to offer a synthesis since no review of the French industrial property market has existed until now; and second, to permit comparison with other European countries. The literature on the subject is limited. An important objective, therefore, is to offer the European reader basic references enabling the subject to be read at greater depth.

The structure and evolution of the industrial property market in France during the 1980s cannot be understood if three factors are not taken into account: the evolution of the structure of the French economy; the specificities of the urban system; and the transformation of the political and administrative system.

The evolution of the structure of the French economy (growth, interest rates, relative positions of economic sectors) has directly influenced the dynamics of companies, and consequently their capacity to develop their properties. The industrial sector accounted for 30.4% of total civil employment in France in 1988 (INSEE, 1990a, p. 97). This sector increased its number of employees between 1960 and 1974 (from 7.004 to 8.307 m), but lost 1.880 m employees between 1974 and 1988. Employment loss accelerated during the period 1980–1988: 0.637 m between 1974 and 1980 and 1.243 m between 1980 and 1988. However, in 1989 there was a gain of 0.03 m industrial jobs. Moreover, industry has experienced a severe decline in net foreign exchange flows during the period 1980–1988, from a surplus of FF 35 milliards to a .loss of FF 43 bn (Dollé, in Boucher, 1989, p .48). These two trends nevertheless contrast 'with evident signs that productive activity is recovering and improving its organization' (El-Rahman, in Boucher, 1989, p. 53), which is clear when considering the new productive investments which are now taking place, the stepping up of productivity, and growth in rates of profit.

The evolution of the industrial sector is very different from the evolution of the other two key sectors, agriculture and services. Agriculture, with 6.7% of total civil employment in 1988, has lost 65.8% of its workforce since 1960 (INSEE, 1900b, p. 81). In contrast, the service sector, with 62.9% of total civil employment in 1988, has seen its labour force grow dramatically during the same period, by 79.3%.

The French urban system (DATAR RECLUS, 1989, p. 21) is dominated by an overwhelming capital, the Paris agglomeration, in which a large part of the population and economic activities is concentrated, and *la province* in which regional agglomerations have played a growing role in the last two decades. The Paris agglomeration, with 9.06 million inhabitants in 1990, dominates the urban structure of France by its size and activities. The other French agglomerations are far behind Paris in terms of population and activities: the biggest are Lyon (1.26 million inhabitants in 1990), Marseilles (1.087 million) and Lille (0.950 million), followed by agglomerations which are of medium size in European terms.

From an industrial point of view, France is divided into two regions along a line from Caen to Grenoble (DATAR RECLUS, 1988, p. 135). The northern region is industrialized, the southern being quite weak. Areas in the northern region have been directly hit by the economic crisis, but there has been a variety of reactions, the most negative ones being in Nord Pas de Calais, Lorraine and Alsace (DATAR RECLUS, 1988, p. 130).

The French political and administrative system is very centralized. The Constitution of 1958 gives full power to the President and the Government, and has generated a powerful bureaucracy at central levels. The Parliament votes laws which are normally translated into practice by the central administration through decrees and orders (Punter, 1989, p. 151–152). Nevertheless, it is important to understand that, in the early 1980s, a strong process of decentralization began, which has given important new responsibilities to regional, departmental and community levels of government, particularly in the economic sphere. The reinforcement of the powers of regional and local authorities during the last decade has given them a major role in the industrial property market. A good example in this respect is the competition between cities to develop science and technology parks (DATAR, 1990a).

The general presentation of the industrial property market will take into account the three dimensions of economic structure and evolution; the urban system; decentralization of the political and administrative system; and will answer three series of questions in sections 5.1–5.4.

1. What is the nature of the industrial property market in France? Who are the suppliers and the demanders? Is it a market of owners or a rental market? What are the quantities of floor spaces exchanged, prices and products?
2. What is the nature of the demand? Which are the legal conditions and financial possibilities which apply to the demander?
3. What is the nature of the supply process? What are the roles of the private and public actors? Particularly, what is the influence of the town planning system?

The characteristics of the industrial property market will be illustrated through three case studies, one in Grenoble, which is an example of a science park, and two in the Paris agglomeration (section 5.5). Vacant industrial land will not be treated specifically in this chapter: this specialized subject is fully reviewed in a study by Lacaze (1985).

5.1 THE INDUSTRIAL PROPERTY MARKET IN FRANCE

The industrial property market is normally presented as a market for 'activities' rather than for 'industry'. This evolution of vocabulary which took place in the 1970s demonstrates that to French ears 'industry' sounds slightly negative, and that floor spaces are now not only 'industrial' but also 'mixed' and organized with services.

The industrial property market in France is segmented: first because of the large variety of industries, and second as a result of the characteristics of the French urban system. It is also a relatively opaque market, for which it is difficult to know real prices. Interesting information has been collected and published by one of the biggest consultancy firms, the Auguste-Thouard group (Auguste-Thouard, 1990a; 1990b). This source has been extensively used throughout this chapter. The market is spatially divided into two: the Paris agglomeration and large French agglomerations (17 in the Auguste-Thouard survey). The two parts are similar in size for floor spaces exchanged in 1989: 1 400 000 m^2 in the Ile de France region and the most lively markets in 1989 measured by transaction volume were those of Lille (262 160 m^2). Lyon (223 500 m^2) and Marseilles (181 000 m^2), which are different in size from other agglomerations.

Most transactions are rental transactions. In the Paris agglomeration, they account for 78.9% of the total, and for around 75% in the other 17 agglomerations. The preference for hiring rather than buying is linked to the high costs of building, mainly in the Paris region, and to the fact that companies are looking for productivity more than for the accumulation of wealth. This is one of the big differences between the industrial property market and the housing market. Three phenomena could account for this characteristic:

1. a preference for plant investment rather than building investment;
2. frequency of change of location because of the development of activities;
3. the need for appropriate buildings for new technologies, since most of the existing buildings offered are inadequate in this respect.

This large majority of rental transactions nevertheless hides a strong preference for buying: outside the Paris region, rental demand is around 55% of total demand, compared with 75% of actual transactions. One of the consequences of this evolution is to encourage the creation of buildings which can be immediately used.

The industrial property market in the Ile de France region in 1989 had the following characteristics:

1. strong dynamics, with an increase of 20% a year in the number of transactions;
2. a widening geographical gap between two areas: the North (and particularly the 'Plaine Saint Denis') and East, with notable growth, and the South and West which remain steady (These movements have been induced by the quick growth of the office market. Paris is handicapped by the level of land prices.);
3. the considerable and general increase in land prices linked with the evolution of the office market, which is progressively pushing traditional industrial activities outside the Paris centre.

The industrial property market in the 17 agglomerations surveyed by the Auguste-Thouard Group is considered as a neglected market, where the floor space exchanged has increased by only 10.6% in 1988/89. Three large agglomerations played a leading role: Lyon, Marseilles and Lille. Lille displays particularly good dynamics, given the perspective of its location in the overall European context, and development in the region, especially of high-speed rail (TGV) infrastructure. Overall, Auguste-Thouard considers the industrial property market is unbalanced, for the following reasons.

1. The supply has too many 'old products' (85%) inadequate to meet demand, the 'new products' offered being too weak.
2. Many industrial estates, most of them derelict, are artificially increasing the total stock on the market.
3. If economic activity rates remain high in the 1990s, many agglomerations will face a shortage of land for industrial uses.

There was a strong reactivation of the industrial property market in the period 1984–1989, with a two-fold increase in building permissions which had to be put alongside the net decrease of loss of jobs in the industrial sector: 247 000 industrial jobs were lost in 1984, compared with 49 000 in 1988. Prices in 1989 reflected a constant rise in prices which started in the mid–1980s in the property markets. Prices are of course very different from one town to another, and from one location to another in the same town. The prices indicated in Tables 5.1 and 5.2 are average values, and should be taken only as indicators. The prices of the Auguste-Thouard group are more precise for locations within towns.

Table 5.1 Prices for industrial buildings (Hors Taxes, 1988)

	Ile de France	Lille	Lyon	Grenoble
Activity parks				
Rents	600–750	350–500	280–350	370
Sales	6500	4000	2800–3200	3810–4200
Activity buildings				
Rents	350–490	230–300	200–340	280–290
Sales	4300	1000	2000–3100	3000
Warehouses				
Rents	250–450	130–180	130–260	200–250
Sales	3200	800	1300–2350	na

Values in francs per square metre.
Source: *Perspectives immobilières*, no. 94, 1989, p. 29

Table 5.2 Prices for industrial buildings (HT, 1989)

	Lille	Lyon	Grenoble
Recent and old			
Centre: rents	200	200–250	na
Centre: sales	1800	2000–3000	na
Periphery: rents	150	150–200	200–250
Periphery: sales	1000	1400–2000	2000–2500
New			
Centre: rents	280	na	na
Centre: sales	3000	na	na
Periphery: rents	220–250	240–300	250–350
Periphery: sales	1800	1800–2800	2500–3500
Top value			
Rents	400	na	na
Sales	3500	na	na

Values in francs per square metre
(Source: *Auguste-Thouard*, 1990b)

Important changes have taken place in the products of the industrial property market between the beginning of the 1970s and the beginning of the 1990s. It is possible to caricature the situation by saying that, at the beginning of the 1970s, industrial buildings were monofunctional buildings located in industrial zones whereas, since the beginning of the 1990s, industrial buildings have been intelligent multipurpose buildings located in green activity parks or, even better, in green science parks.

Flexibility and adaptability are the key words for industrial buildings. Why? Mainly because growing European and international competition has increased the speed of change in the economy. Developments remain viable if they remain attractive to as wide a range of potential occupiers as possible in this rapidly changing environment. But it could also be argued that the French economy has seen significant restructuring during the period. The evolution in the quality of buildings in that sense reflects the economic evolution, particularly the increase in the needs of the private sector. This has caused development of shared buildings to appear, combining production with stocking and service activities. The floor space per job has also increased, from about 10 m^2 to 20–25 m^2 today.

At the beginning of the 1990s, then, there was a tremendous development of multipurpose buildings with an immediate opportunity to connect with associated offices, communication systems and networks. There is even doubt about the use of the words 'industrial building' because of the flexibility of the buildings and their adaptability to activities other than industry.

A pronounced change has occurred in attitudes, across the spectrum from a production to a marketing point of view, including the role of the building itself in contributing to the well-being of employees, which has been recognized as an important factor in production and in the company image. It is not only the building which has evolved; it is also its location. Important aspects are site and environmental quality, and transportation systems (motorways, roads, railways, airports). These also contribute to the preference for flexibility, and therefore the choice of renting rather than buying.

5.2 LEGAL, FINANCIAL AND TAXATION ASPECTS

5.2.1 Acquisition of industrial buildings

The user of an industrial building can be an owner or a lessee. The choice between the two depends on the firm's economic strategy, and in particular the time perspective of its business plan.

A company which wants to have industrial buildings has several choices, from short-term renting to buying. In each case there is a specific legal, financial and taxation system which applies, and which can only be summarized here. The main legal and financial possibilities are the following: lease of less than two years; simple lease (*bail commercial*); *crédit-bail;* selling (buying) by payments by instalments; *bail emphytéotique;* construction lease (*bail à construction*); selling (buying) property.

Lease of less than two years

Leases of less than two years take the form of a written 'convention' between the owner and the lessee. The application is defined by the *code civil* (Article 1712 and following). This procedure is also called *usine-relais*, and is frequently used in 'company nurseries', designed to assist the start-up of new enterprises. It is attractive for new companies with good development prospects, and the need to move quickly as they expand. The problem for the companies is that they have no entitlement to a conventional renewal. The owner of the land is able to choose another tenant company. For the municipality there is also a risk that the company cannot pay the rent, but prevents the acquisition of a new lessee while it remains. This procedure is therefore much less used than the simple lease, which is the most common system in France.

Simple lease (bail commercial)

The *bail commercial* is a written contract signed by the owner and the lessee for a minimum period of nine years. It has to follow legal rules on *baux commerciaux* as defined by *décret* no. 53–960, 30 September 1953. The decree applies to all property for commercial, industrial or handicraft uses. To have access to that category of contract the manufacturer has to be registered in the *Registre de Commerce* of the Chamber for Commerce and Industry. The renewal of the *bail commercial* is nearly always automatic, unless the owner is giving compensation. The lessee has a right to terminate the contract every three years, without any allowance, independently of the duration of the contract.

The simple lease (*bail commercial*) has many advantages for the lessee, the first being the stability of its situation. For a detailed discussion, see Thouard (1989a, pp. 281–294). If there is any problem in the use of this category of lease, the *Tribunal de Commerce* (Commercial Court) is competent to test the legality.

Crédit-bail

The *crédit-bail* is a financing method by which a lender offers the lease of a property to a borrower, combined with a unilateral promise of selling (*promesse unilatérale de vente*), applied at the end of the financing period by a transfer of property from the owner to the borrower (Thouard, 1989a, p. 301). This hire-purchase method has been defined by law no. 66–455 of 2 July 1966 on *crédit-bail* and ordinance no. 67–837 of 28 September 1967 on *sociétés immobilières pour la commerce et l'industrie*. The duration of the *crédit-bail* is between 15 and 20 years, unless the borrower wants to end it, in which case he has to pay a termination indemnity.

The *sociétés immobilières pour la commerce et l'industrie* (SICOMI) are specialized companies which provide *crédit-bail*. The SICOMI have an investment role by buying and letting properties. As there is a large number of SICOMIs, the forms of *crédit-bail* are very diverse, with differences in interest rates, duration, residual value, indexation, and also mode of payment, which can be progressive or regressive.

The *crédit-bail* was attractive for companies at the end of the 1980s with the decrease of interest rates, and because it provides a property at the end of the 15–20 years. Also, the company has no initial capital to provide, since financing covers 100% of the property costs. The main problem for companies could be the duration of the contract, because it is less flexible than the simple lease. This is why it is necessary to allow for ways of reducing the contract duration. Two possibilities exist: transfer to another user, and buying of the property by the company. The latter is the most difficult because of the funding required.

Selling (buying) by payments in instalments

The advantages of the *crédit-bail* also apply to another possibility: selling by payments in instalments. The property transfer and the payment are not concomitant. It is an ordinary sale, with immediate transfer of property to the manufacturer with facilities for payment. This solution is dangerous for the seller because he loses all his property rights and has little chance of recovering his assets if the company collapses.

Bail emphytéotique

The *bail-emphytéotique* is a long-term lease (99 years) which gives the taker absolute rights to the property in exchange for a light indemnity.

At the end of the lease the lessee becomes the owner. This is used in the agricultural sector (rural code articles 937 to 950) and is not really adapted to the industrial sector. Industrial activities are too varied for this sort of lease to be appropriate, and it has been replaced by the *bail à construction*.

Construction lease (bail à construction)

The construction lease (*bail à construction*) is a contract for 20 years minimum, which allows the lessee to build on the land of the owner. The lessee has property rights on the buildings. The lessee is given all possibilities of selling and constructing. The lease defines all relations between the lessor and the lessee. It has to follow the rules of law no. 64–1247 of 16 December 1964 (construction code, articles L 251–1 to L 251–9).

5.2.2. Selling (buying) property

Selling (buying) property is a property transfer by a deed between an owner and a taker. That deed is subject to numerous laws and taxation rules. In general the owner and the taker sign, before the deed itself, a sale promise (*promesse de vente*) which defines precisely the terms and clauses of the property transfer. This also offers the opportunity to fulfil administrative requirements, which are numerous and can take a long time, as can the delivery of the *certificat d'urbanisme*. The most common form of *promesse de vente* is the unilateral promise of selling (*la promesse unilatérale de vente*), by which a seller says that he is selling his property for a defined price to a defined person or company. If the buyer accepts, implementation of the sale will depend on the conditions which have been accepted by the parties concerned; dates, indemnities for immobilization of capital, conditions under which the sale may be suspended.

Taxation system

There are three categories of taxation system linked with the selling of industrial property: *enregistrement*, (registration), *taxe á la valeur ajoutée*, TVA (value added tax) and *plues-values*.

The registration tax (*taxe d'enregistrement*) is the transcription of the new rights linked with the transfer of property. It is a tax defined by the General Code of Taxes. It is not used when TVA is applied; in this case there is only at *taxe de publicité foncière* (0.6%). TVA applies to the selling of land and property. It is paid by the seller at the rate of 18.6% of the transaction cost or the actual cost of the building.

The increase of value between two selling transactions is subject to a special tax system by law no. 76–660 of 19 July 1976. The tax is owed by individuals and associations of persons, on land and buildings, when they pay income tax only. The taxation rate varies according to the duration between the two transactions. This tax system is heavy compared with most other European countries.

Brokerage fees

Brokerage fees vary greatly from one country to another in Europe; France has one of the highest. The transaction costs for selling add up to 25% (legal 1%, brokerage 5%, stamp duty 19%) plus 18.6% TVA on brokerage fees. Transaction costs for letting are 17.6% (legal 1%, stamp duty 16.6%) plus 30% of first year rent. TVA at 18.6% has to be added to all fees, plus a 2.5% local tax.

Hazardous installations (installations classées) and environmental protection

Classified installations (law no. 76–663 of 19 July 1976) are all activities (particular industries) which may generate danger or inconvenience to their environment. Plants are categorized by decree in three classes, from the most dangerous to those that cause only inconvenience. They are submitted to Prefect authorization (first and second classes) or to declaration (third class). The authorization application needs an environmental impact study, and is submitted to public inquiry (Thouard, 1989a). The declaration is a formal act which informs the Prefect. It is automatically approved unless the Prefect wants to change the project for legal reasons.

5.2.3 Financial incentives for activity location

The French government provides grants or financial incentives for activities, particularly in the industrial sector, which fulfil national spatial policy. This policy of *aménagement du territoire* has been in force for the last 25 years and applies to the regions of the West, the South West, the Massif Central and the industrial conversion zones (including enterprise zones).

Financial incentives are mainly grants. In the last 25 years, about

F 25 bn at 1989 values have been used and 750 000 jobs supported by different investments (DATAR, 1990b, p. 109). The main forms of assistance and incentive are as follows (DATAR, 1988, 1990b; Thouard, 1989a):

1. the National Spatial Planning Premium (*Prime d'Aménagement du Territoire*), which is the most important was simplified in 1986 and takes the form of an equipment grant given to support job creation;
2. temporary exemption from professional tax;
3. financial incentives by conversion companies (about FF 400M a year);
4. enterprise zones created on three sites, with a complete exemption from society tax for 10 years (Heurteux, 1990);
5. regional financial incentives which are variable, and a function of the economic policy of each regional council.

5.3 PRIVATE AND QUASI-PUBLIC BODIES

The private and quasi-public actors have changed in the 1980s. At the beginning of the 1980s the market was weak, and even declining. In 1985–86 a new period began, with a large increase in the number of actors and the arrival of new investors.

It is difficult nowadays to define professional boundaries precisely: there is no clear definition of the role of each actor in the process of development. Nevertheless it is possible to identify what the actors are doing in that process. There are six main categories: occupiers (owners or lessees); *aménageurs* (who equip land with infrastructure facilities); developers (*promoteurs*) (who coordinate the construction process); builders; investors (who finance the development and selling of properties); consultancy companies on property (who counsel clients and sell buildings).

5.3.1 Occupiers (owners or lessees)

Until the 1970s, the owner-occupier was dominant on the market (Auguste-Thouard, 1988). The company was acting alone, looking for land on an appropriate site, buying land and building. That situation evolved as renting became more and more common. Property ownership came to be seen as a constraint on a company's freedom of action, and possibly a bad investment as the economy became more dominated by the service sector. The impact of that evolution at the end of the 1980s led to the construction of buildings without knowing the buyer (*l'immeuble en blanc*). The owner is not necessarily the occupier.

5.3.2 *Aménageurs*

The *aménageur* is a body in charge of transforming the land into building land by providing all the necessary infrastructure. It may or not purchase the land before realizing works. It could be a public, quasi-public or private body. If we disregard the few individuals who undertake the servicing of their land themselves, the *aménageur* role in France is mainly undertaken by local authorities, which are studied below. Private *aménageurs* could play an important role, but the quasi-public bodies are fundamental, particularly the *Caisse des Dépôts and Consignations* group which will be studied alongside developers (*promoteurs*).

5.3.3 Developers (*promoteurs*)

The main activity of developers (*promoteurs*) is the coordination of construction or renewal. They can also play an integrative role from the purchase of vacant land to the selling of buildings, with the equipment of land and the construction of buildings. They do not always know the users.

The industrial property market works less and less on the basis of the traditional relationship between developer and client, and more and more on the triangular relationship investor-user-developer (Leloup, 1986). The consequence has been a change in the way development companies act, because in general they do not have enough funds to undertake the risk of the project. Developers are increasingly looking for builders or investors; they offer them their know-how in managing and coordinating property development projects.

The *Caisse des Dépôts et Consignations* group has a sound knowledge of property development (including industrial property development): this group has been predominant since World War II in implementing industrial zones. It has developed specialized tools for local authorities. This huge activity is based on the creation of numerous subsidiary companies of the group: the Central Company for Territory Equipment (*Société Centrale pour l'Equipment du Territoire*, SCET); mixed public private development companies (*société d'économie mixte*, SEM); development and selling companies (*Société de Promotion et de Commercialisation*, SOPREC); and the Central Company for Property (*Société Centrale Immobilière de la Caisse des Dépôts et Consignations*, SCIC) which has realized a large number of buildings for the main French administrations. In order to respond to the growing demand for new 'products' with new technologies, the SCIC has created a new specialized company, G31 (Investment in Industrial and Shopping Property Group) (SCIC, 1988).

G31 is acting from three different points of view: first, it identifies for local authorities the most interesting sites for developers, building contractors and investors; second, it can be developer (*promoteur*); and, third, building contractor. The advantages of the G31 company are important for local authorities: it is a subsidiary company of the *Caisse des Dépôts et Consignations* and as such has excellent financial standing; and it has the know-how of all the other subsidiary companies of the group. Moreover, the *Caisse de Dépôts et Consignations* has created a specialized company called ASSURECO for the mutualization of risks, its purpose being to cover risks taken by local authorities when they buy industrial buildings for private companies, risks of non-payment of rents, and risks linked with borrowing guarantees given for development of activity zones or properties.

5.3.4 Builders

The builder carries out the work of constructing buildings. He could also be the developer. Large building companies have appeared on the property market in the 1980s, as a consequence of their renewed growth in France. Their substantial liquidity has given them the opportunity of taking on the risks of the property market. These companies have also created their own development companies.

5.3.5 Investors

Investors are insurance companies, banks, pension funds, and specific financial institutions (*société immobilière pour le commerce et l'industrie* and *société civile de placement immobilier*). At the beginning of the 1980s, there were only a few property investors, in comparison with the demand from companies (ARU, 1985). The weakness of the investment on offer was due to the general economic and political context and also to changes in the way funds were collected and used (capitalization in the Paris stock exchange). That the situation was amplified by other cyclical factors, such as economic uncertainty, and the costs of borrowing.

In the mid–1980s, with the economy expanding, new investors have collected more money and have injected it into the property market. The institutional investor became a new partner on the property market. Two institutional arrangements have developed extensively, the *sociétés civiles de placement immobilier*, and the *sociétés immobilières pour le commerce et l'industrie*.

Société civiles de placement immobilier (SCPI)

The civil companies for property investment (*sociétés civiles de placement immobilier*, SCPI) are property companies which publicly collect savings and use these funds for property investments. The market was born in the 1960s on the initiative of the property profession and has been regulated by the law of 1970, under the control of the Paris Stock Exchange Commission. Since the beginning of the 1970s the commercial development of these companies has been steady: they are now very powerful on the property investment market. From 1976 to 1984, savings of around FF 1 bn were collected every year. In 1985, this figure rose to FF 1.7 bn; in 1986 it reached FF 5 bn; and in 1987/88 the figure was in the region of FF 8–10 bn p.a.

The number of SCPIs specializing in property for companies is about 120, which means a capitalization of about FF 30 bn (by comparison the insurance companies have a capitalization of FF 100 bn). The geographical distribution of investment is 60% in the Paris region and 40% elsewhere. The allocation by sector is: 40% office property; 40% industrial property; 20% shopping property.

An institutional investor on the industrial property market could collect savings by creating a *société civile de placement immobilier*. This will offer returns of 4% to 5% on housing investments, 6.5% on office and industry investment. The developers sell buildings to the SCPI with a guarantee of 9% return a year, on the basis of rents or *crédit-bail*.

Sociétés immobilières pour le commerce et l'industrie (SICOMI)

The property companies for commerce and industry (*sociétés immobilières pour le commerce et l'industrie*, SICOMI) were created in 1967. Their purpose is to let office or industrial properties through the simple lease (*bail commercial* 3–6–9) or the property *crédit-bail*. It is possible to created *crédit-bail* outside a *société immobilière pour le commerce et l'industrie* but in this case there is no tax advantage (exemption from society tax, discharge rules).

Two thirds of the investment of the SICOMIs are in the Paris region, and there is a distinct lack of information about the markets outside the Paris region. Their criterion for intervention is security, and they invest predominantly in large agglomerations. In this way, if any problem arises over the company renting the property, it is easier to find a new tenant. SICOMIs are also more interested in the rental value of the building than its market value. They have been more directly interested in office property. Their preference for industrial property favours buildings that can be easily divided.

The SICOMIs are increasingly dominant on the property market: 52% of investment (FF 22 bn) in 1988 as opposed to 43% in 1985. The SCPIs have decreased in importance: 15% of investments in 1988, 19% in 1987. The share of pension funds and insurance companies has also decreased: In 1985, 22% of investment declining to 18% in 1988.

The arrival of foreign investors (particularly the Japanese) was very important at the end of the 1980s. In 1985 foreign investment was 5% of the total. In 1988 it represented 10% with FF 4.5 bn, including FF 3 bn for the Japanese. After the United Kingdom, France is now the second market in Europe for Japanese investments.

5.3.6 Property consultancy companies

Property consultancy companies are professional advisers which commercialize existing properties and prospect for new developments. Consultancy firms play an important role in the process of property development, with an increase in the number of transactions. They have taken the commercialization role, replacing the developers and promoting their own activities. The structure of this sector of activity is characterized by concentration: of about 40 companies on the property market, the first ten companies account for 50% of the transactions. It is a highly competitive market. With the challenge of 1993, and severe competition from British consultancy firms, which are still leaders in the market, a strategy of opening up foreign markets has taken place, in association with other large companies or the creation of subsidiary companies.

The consultancy profession has transformed itself fundamentally in the last decade. Until World War II they were not selling any new properties, only old ones after improvement. The evolution up to today has been due to the transformation of the French economy, and competition with British consultancy companies, which have created this new activity. Nowadays we can speak about 'property engineering', an activity that spans from the very beginning to the selling or letting of property.

The main activity of consultancy companies is to provide a commercial property service according to the needs of their clients (they are well known for that activity), from the elaboration of the project to its management. A consultancy company identifies appropriate sites for its clients and proposes them to developers with whom the project is analysed. Local authorities sometimes have their own bodies for commercial property management but they also work with the consultancy firms (as in the case of ZIRST, Grenoble: see case study below) when there is demand from companies. In this case the consultancy company is acting on the behalf of the local authority, but with a certain degree of independence from it. The consultancy company can act early in the processes of development and construction. Many companies are not well organized in this area, and the professional advisers could anticipate the emergence of the demand.

Nowadays there are very many actors in the property development process and they compete fiercely. There are many challenges, notably that of 1993.

5.4 PUBLIC ACTORS AND TOWN PLANNING

In the early 1980s, a strong process of decentralization began, which gave important new responsibilities to regional, *département* and communal authorities, particularly in the economic sphere. At the end of the 1980s local authorities had a powerful role in the industrial property development process, and central government did not intervene very much, except for very big projects, and in problems of dereliction.

5.4.1 The economic intervention role of local authorities

With decentralization and the economic crisis, local authorities as a whole have taken possession of their responsibilities in the economic sphere, and their attitude has changed deeply from the one they had in the 1970s. Local economic development has become a fight against unemployment, and the elected representatives have changed their role in consequence: in addition to their traditional role in delivering public services, they now act as public spatial developers (Bouinot, 1987). These roles are linked together: the economic strength of a commune has direct influence on its possibilities for financing public equipment, particularly because industry is providing the professional tax, a very important local tax. The general trend is, therefore, a change from a land use policy to one of friendly reception of companies. Local authorities increasingly take a more global and qualitative approach to local economic structure, and they have developed their own administrations for economic development. These administrations collect information on the town economy, promote the town economic activities, try to attract new companies and fight for developing employment. At *département* level, this role is often assigned to *comités d'expansion économique*.

The financial incentives and grants that local authorities can give for industrial property development, after the decentralization process, can be direct or indirect. Direct assistance can take the form of loans, cash advances, and interest allowances. Forms of indirect assistance include loan guarantees (with limitations), and discount on rents and the market value of industrial properties.

Local authorities can purchase industrial buildings and sell them after renewal but with a maximum discount. This is the difference between market price and cost price. Nevertheless, in National Spatial Planning Premium zones (*Prime d'Aménagement du Territoire*) local authorities could give a 25% discount supplement on the property market price or rents corresponding to that price. (*La Gazette des Communes*, 1 November 1987).

Local authorities can also give temporary exemptions from professional tax and other local taxes.

Another way of supporting companies is to help them with their initial capital and working capital. This has the advantage of really helping small and medium-sized companies at the very beginning when their profitability is low and banks are reluctant to finance them. Local authorities can play a very important role in those financial aspects, particularly because the borrowing capacity of small companies is very limited at the beginning (Pecqueur, 1989). This is one of the reasons why there has been a development of ways of collecting savings at local level to promote companies, because in this situation borrowers and lenders are in close proximity; the lenders know the users of their funds, and the interest of the company project. Three ways of collecting money are used: local guarantee funds; investment club networks; capital collecting. For managing these funds, local partners can use several tools: associations, limited companies, *fonds communes de placement à risques* (created by law of 3 January 1983), *capital-risque* companies (created by law of 11 July 1985) and the *société d'économie mixte*.

In 1987, economic interventions of local authorities were characterized by:

1. the predominance of communes in the economic assistance (47.4% of total assistance);
2. diversification of the commune assistances which were not only 'traditional' (grants, premiums, loans, advances);
3. priority given to economic development (98.1% of the total).

The traditional assistance given by local authorities has been steady, with a huge increase in discounts on property market prices and in rents for land and buildings: FF 89.3m in 1986, rising to FF 420m in 1987.

Analysis of local authorities by category shows that in 1988 commune assistance amounted to FF 4.1 bn, resulting primarily from the actions of large communes. This comprises mainly loans and advances to companies (20.1%), grants (16.1%), and industrial zone development (12.1%). This last increased by 37% in comparison with 1986, and is a significant indicator of the intervention of municipalities in infrastructure development. Assistance at the *département* level was F 2.1 bn. This included a decrease in the predominant role of direct assistance (from 79.9% in 1986 to 73.8% in 1987) due to the strong increase of indirect assistance (+75.6%) and particularly discounts on property and land market prices and rents, which increased from FF 22.4m in 1986 to FF 29.01m in 1987. This category of assistance represented 13.6% of the total, but it was mainly directed towards housing. The situation was similar with loan guarantees. Regional assistance totalled FF 2.4 bn.This was mainly directed towards economic development (99.5%). Grants and premiums were the main form of intervention by the regions (80.5% of the total).

5.4.2 Urbanism documents and rules

The main urbanism documents are the *schéma directeur* and the *plan d'occupation des sols*. The main urbanism procedures are the *permis de construire*, the *lotissement* and the *zone d'aménagement concerté*. Only a short introduction is possible here. For a fuller account, see Punter (1989).

Urbanism documents

The *schéma directeur* (Urbanisme Code, articles L 122–1 and following) defines the main and long-term orientations of spatial development of a territory. It has to combine housing, economic and transportation development, and conservation of the environment. It has not been considered a very useful document since 1967. It applies only to administrations, not to individuals or companies.

The *plan d'occupation des sols* (Urbanisme Code, articles L 123–1 and following) is the most important urbanism document: it gives construction rights on specified areas, and defines urban developments. Its importance has been reinforced by the decentralization process: if a commune has a *plan d'occupation des sols* it can be autonomous in its spatial development.

These documents can introduce problems for industrial activities, and many argue that they hinder development. In fact the flexibility, as a result of this decentralization, is actually very important and may be exploited by the mayor and the commune (Sauvez, 1989, pp. 135–137). Nevertheless, problems can arise if there is competition between industry and other categories of activities, particularly housing or agriculture.

The *plan d'occupation des sols* gives a general definition of the areas in a town which should be used for industrial activities. But it is only a first step towards development: the creation of a building or of an activity area needs operational tools and urbanization procedures which are more effective and can transform ordinary land into an industrial estate.

Urbanism procedures

The general urbanism procedure can be used to create industrial properties or to develop activity zones corresponding to a voluntary policy of grouped development of industries on a site. To permit the implementation of infrastructure on land, three procedures are very important: two have a private nature (the *permis de construire* and the *lotissement*), while one is of initiative and public interest: (the *zone d'aménagement concerté* (ZAC) procedure).

The building authorization (*permis de construire*) is an authorization given by an administrative authority to build one or more new constructions, or to change existing buildings. Obtaining this authorization is compulsory for any building. For industrial buildings, the authority could demand works to provide roads, water supply, sewage system, lighting of the area, car parking or environmental areas. These have to be provided out of the developer's own budget.

The *lotissement* divides land prior to building on it. Buildings can be industrial. The person who benefits from the *lotissement* authorization has an imperative delay (from 18 months to three years) to complete all infrastructure works on it before selling it. The *lotissement* is of significance when one knows what to do on a plot (share number and map): the *lotissement* regulations are very precise, which could be a constraint for *aménageurs*. From a developer's point of view the interest is that there is no public enquiry or environmental impact study before completing infrastructure works. As soon as there is division of a site into three parts, it becomes necessary to implement a *lotissement*. The building authorizations are given when shares are sold, for construction. They are given according to the *lotissement* and *plan d'occupation des sols* regulations.

Numerous activity zones have been developed under the *zone d'aménagement concerté* procedure. The Urbanisme Codes says (art. L 311–1 and following) that it is a zone in which a competent local authority or public body (*établissement public*) decides to intervene in order to implement infrastructure works on land units which have been bought, in order to sell them to public or private users.

The *zone d'aménagement concerté* needs the elaboration of a brief with the spatial zone limits. An environmental impact study is necessary, when there is no *plan d'occupation des sols*. The definition of the implementation (legal and financial) procedure is compulsory, and so is the preparation of a more specialized plan (the *plan d'aménagement de zone*), which replaces the *plan d'occupation des sols*. This specialized plan includes all the development aspects, with construction rights and the *coefficient d'occupation des sols* (plot ratio).

The main bodies who have implemented activity zones in France are local authorities or private developers (*promoteurs*). Local authorities have all the legal rights to initiate the *zone d'aménagement concerté* procedure. Nevertheless, raising the necessary finance is sometimes very onerous for them, and there is also a lack of officials who are technically competent in real estate development. Therefore, they create mixed public-private companies (*société d'économie mixte*). Many of these are subsidiary companies of the Central Company for Territory Equipment (*Société Centrale pour l'Equipement du Territoire*, ZCET), itself a subsidiary of the *Caisse des Dépôts et Consignations*.

When a *zone d'aménagement concerté* is implemented by a private developer the financial aspects of the brief are different from the public *zone d'aménagement concerté*. In this case there is no possibility of getting grants for development, and the investment is thus of the local authorities and/or the private developers. Financing an activity zone can be a major problem, mainly due to the time needed to implement the necessary infrastructure. Special grants (with low interest rates) from the *Caisse des Dépôts et Consignations* are usually used.

The *zone d'aménagement concerté* can be implemented by local authorities in a number of different ways. If they do everything themselves, using their administration, it is known as a *régie* case. It can also be convenient for them to give the process to an *aménageur* by one of two procedures: a convention or a concession. In the former, the local authority gives permission to a quasi-public or private body to realize the infrastructure works and eventually the buildings. This procedure is called private *zone d'aménagement concerté* because it is realized at the financial risk of the quasi-public or private body. The second possibility is the *concession d'aménagement*, in which the local authority gives a company, under its control, responsibility for infrastructure works and equipment. The local authority has the financial responsibility; the result is called a public *zone d'aménagement concerté.*

The *zone d'aménagement concerté* procedure is used for large projects in which it is necessary to know only the general orientations of the development. The *plan d'aménagement de zone* can be changed, and this is an important difference from the *lotissement* procedure. It is one of the reasons why the replacement of a *plan d'occupation des sols* by a *plan d'aménagement de zone* needs a public enquiry.

5.5 CASE STUDIES

The case studies below have been chosen from opposite ends of the spectrum of possible examples to show the processes at work in the industrial property market in France; they combine most of the different aspects described separately above. At one end of the spectrum, there is derelict industrial land, which is often the most difficult case, and which has been the object of specialized policies from the State (Lacaze, 1985). In this category, two case studies in the Paris region are presented: one was elaborated and implemented through private companies, the other by a commune. At the other end of the spectrum, we will present a development for high technologies, which has been the object of vigorous policies both at the national and local level in France, under the label *technopole* (Bruaht, 1990). There were about 20 *technopoles* in 1990. The Grenoble experience is one of the first; it was developed by local authorities, which is interesting in a time of devolution of power to local levels of government.

5.5.1 Case studies in the Paris region

The EVOLIC activity centre

The EVOLIC activity centre in the Gennevilliers Commune (*Hauts-de-Seine*) was implemented on 3 ha of derelict industrial land. (IAURIF, 1986, pp. 101–104). The project was to build an activity centre with industrial buildings. This centre was developed by private developers and investors, who had purchased the land from the Commune. The derelict land was purchased by the *Compagnie Générale pour l'Immobilier de l'Company* (CGI), which acted on behalf of the *Caisses Centrales de Mutuelles Agricoles* (which invested about FF 144m). The CGI was responsible for the development process and for definition of the construction programme (choosing in particular as building contractor an architectural office, Le Grand Rabinel). Three buildings were constructed, of 23 128 m^2, with offices (30%) and activities (70%). The floor space for each of the premises was between 67 and 2000 m^2; the average floor space was 400–500 m^2. The premises were let under the *bail commercial* system; the average renting price was between FF 327 – 420/m^2. The management and marketing were given to a *société immobilière pour le commerce et l'industrie* which was a subsidiary company of the CGI.

The location of this centre (west Paris, with good links to Paris centre, La Défense, and industrial zones) explains the rapid take-up of premises between January 1984 and March 1985.

DUBE Company rehabilitation

DUBE Company (previously a welding company) were completely rehabilitated by the Plaisir Commune, Yvelines, west of Paris (IAURIF, 1986, pp. 82–86). The floor space of the premises was 1268 m^2 on a site of 400 m^2. The project was to define new uses for a vacant building in an industrial zone with no vacancy (30 companies with 600 to 700 jobs). The building itself was in a very poor state, and could not be left in those conditions without ruining the image of the whole industrial zone. Furthermore, the idea of the Commune was to encourage the arrival of medium-sized industries or companies, which could develop on the new *zone d'améngement concerté* promoted at the same time.

The development project began with the purchase of the buildings by the Plaisir Commune in July 1985, for FF 2.8m (excluding fees), FF 700/m^2, with a loan from the *Caisse des Dépôts et Consignations* (10.5%, redeemed over a 15-year period). The development project and construction of the buildings were the complete responsibility of the technical administration of the Commune. Works were realized with a loan from the *Caisse d'Aide à l'Equipement des Collectivités Locales*, a bank which has close links with the *Caisse des Dépôts et Consignations* at 12%, redeemed over a 15-year period. The buildings were managed directly by the Commune, under the *régie* system, as was the letting of the premises. The Chamber of Commerce contributed to the marketing of the project by distributing information.

The area was defined as a nursery company zone for new technologies, with twelve shares. Three categories of premises were offered: production workshop (172 – 365 m^2; offices (16.5 – 70 m^2); warehouses (all in all 335 m^2). The lease was under the short period system (less than 2 years) at the following prices: workshops FF 240 – 300/m^2 per year, VAT included, without working expenses; offices FF 480 – 520/m^2.

5.5.2 The Grenoble agglomeration and the ZIRST of Meylan

The Grenoble agglomeration, with 400 000 inhabitants in 1990, is the tenth largest French agglomeration, and a medium-sized agglomeration at European level. Grenoble has experienced good economic dynamics in the last two decades, based on an industrial tradition. The 1980s have seen the economic crisis, with a strong process of recovery at the end of the decade. In the last two decades the agglomeration had developed a specific link between industries and universities, based on research activities which have an international reputation in many sectors (DATAR, 1989).

The Grenoble agglomeration is located in a powerful French region, the Rhône-Alpes region, and a prosperous *département*, Isère. The Rhône-Alpes region, second in size after the Paris region, has used the economic crisis of the early 1980s to adapt the composition of its industries and services, and has improved its development at the end of the 1980s: the economic growth of the region was higher than the French average from 1986 to 1989 (*Le Monde*, 17–18 1991, p. 16), and population and employment grew in *Isère* as well as in the region as a whole. Local authorities play an important economic role in the Rhône-Alpes region and in the Isère *département*. In the Isère *département* particularly, local authorities have for a decade operated an assistance system for companies, which focuses on land and buildings, in order to support employment. This assistance is both indirect (loan guarantees, discount on rents and market value of the industrial properties) and direct (loans, advances, interest allowances).

Most local authority actions in the Isère *département* are land purchases and discounts on the market value of industrial properties (62% of the total). Assistance from communes predominates (40% of the total), especially for land purchases (93%). These actions are sometimes on the edge of legality: for example, the discounts on rents or market values of properties are accepted only up to a 25% limit inside the special development areas, as classified under *aménagement du territoire* procedures. In the Isère *département*, FF 4 475 902 were spent in indirect assistance in 1985, FF 18 437 336 in 1986 and FF 27 328 600 in 1987.

The industrial property market of the Grenoble agglomeration

The supply of industrial buildings in 1987 was steady in comparison with that of high technology business space, which was booming; the high technology business space market multiplied 2.5 times and it represents 90% of the projects. The share of new buildings has risen (+16% of annual growth). That growth has a strong influence on floor space (which has risen too) and the legal use of buildings (diversification). The average building on the market is an old building, which is free for rent, with a floor area between 500 and 1000 m².

Renting possibilities developed strongly at the end of the 1980s:

	Industrial buildings	*Multipurpose buildings*
1986	60%	82%
1988	78%	84%

The sale of industrial buildings rose strongly in the mid 1980s, from 6% in 1985 to 40% in 1986, mainly for important areas (more than 2000 m²).

The Grenoble industrial property market in 1989, from the point of view of the Chamber of Commerce, was relatively short of multipurpose buildings, even if the total stock of industrial buildings rose, due to industrial changes. Letting of industrial buildings decreased in 1989 by comparison with 1988: 38 500 m² let in 1989 (including 10 600 m² in the town centre), and 66 224 m² in 1988. The rise in demand has not been satisfied because of the lack of land reserves in the periphery of the agglomeration. Moreover, land prices are too high for many companies, and most existing buildings are inappropriate and too old. The Communes also have a tendency to reject heavy industries in order to attract light activities, especially in high technologies, with good financial guarantees.

Rents in the Grenoble agglomeration are a function of the category of building (age, area and size) that companies are looking for. Prices in 1989 were the following (in FF; ht signifies *hors taxes*, without tax):

offices	FF450–500 ht/m^2
activity buildings:	FF280–300 ht/m^2
warehouses:	FF230–280 ht/m^2

The industrial property stock in the Grenoble agglomeration is getting older, and many of the buildings will have to be completely rebuilt. There is a strong demand for new medium-sized buildings (between 80 and 100 m^2).

The zone d'innovation et de recherche scientifique, Meylan

The idea of the *zone pour l'innovation et les réalisations scientifiques et techniques* (ZIRST) was conceived at the very beginning of the 1970s, with the objective of promoting the creation of high-technology innovative products and the interface between industry, research and university. The choice of the Meylan *commune* was the result of the priority given to environmental quality, in a residential suburb located just by the University. That zone, which was initiated by forces of civil society, was developed with strong local authority policies, without assistance from central government.

At the beginning of 1990, there are 200 companies in the ZIRST, with 5000 jobs in 110 ha (including 45 free hectares in the Montbonnot *commune*). To qualify to operate in the ZIRST it is necessary to apply to a special management committee, which is in charge of the implementation of the objectives of the activities. Most companies on the site were created in the park itself (60%) or are divisions or subsidiaries of large companies (39%). Industrial users (excluding services) include; artificial intelligence (3%); robotics (5%); remote and off-line process control (6%); automation (7%); training (11%); logiciels (software) (13%); industrial electronics (17%); sensors (components and complete equipment) (17%); computers and peripheral activities (21%).

How was this activity zone implemented? After the development of the idea at the end of the 1960s, a *zone d'aménagement différé* (ZAD), a category of ZAC, of 200 ha was created in 1971 to prevent any speculation on land prices. The area was developed from 1971 under

PARC DE

Figures 5.1–5.4 Development in ZIRST, Grenoble, France.

the *zone d'aménagement concerté* procedure, with a concession to the *département* mixed public-private company, the *Société d'Aménagement et de Développement de l'Isère* (SADI). The area was classified as an activity zone in the *Schéma Directeur d'Aménagement et d'Urbanisme* in 1973, and defined in the *Plan d'Occupation des Sols* of Meylan as a future urbanized zone (100 ha, *coefficient d'occupation des sols* = 0).

The *Société d'Aménagement et de Développement de l'Isère* became the *aménageur* and the builder, with reversion of buildings and facilities at the end of a 15-year period. Responsibility for the promotion of the zone was assumed by a specialized structure, the PROZIRST company, with the help of public and private actors (Chamber of Commerce, banks). PROZIRST is a limited company, created in 1972 to promote the zone, but also to pool resources in the site, to organize leisure activities and to find solutions to problems that confront all companies.

A new experience for developing new industries started in the 1970s with the creation of buildings for renting, for periods of less than two years. The LOCAZIRST are Commune of Meylan properties, with a management mandate given in 1986 to a specialized company, *France-Régie*. This company is in charge of commercializing and managing four zones of 300 m^2 (12 000 m^2 overall).

The *zone pour l'innovation et les réalisations scientifiques et techniques* was developed at the end of the 1980s on the Montbonnot Commune. The *Société d'Aménagement et de Développment de l'Isère* was reinforced by a private company, the Michel Ferrier group. The construction of buildings was given to private *promoteurs*. A civil company was created for each property development project, in association with the Michel Ferrier group and banks. The Michel Ferrier group is also responsible for the creation of a business and service centre. Letting was assigned to the national consultancy companies Auguste-Thouard, Jean Thouard, Bourdet, and SODEC (a subsidiary company of the *Caisse des Dépôts et Consignations*).

5.6 CONCLUSION

The industrial property market in France changed dramatically during the 1980s, and one of the most important evolutions has been the local authorities' attitude towards development. The land-use planning instruments are, in this respect, tools which are used to support a local economic policy aimed at attracting companies. The activity zones which have been created are not only development areas, but are also areas in which *communes* welcome companies in solving their global problems, including such aspects as leisure activities for employees. The rise of environmental concern has also favoured new-technology companies, which have a good image in this respect.

REFERENCES AND FURTHER READING

Abd-El-Rahman (1989) in Boucher, M. p. 53.

ADEF (1989) *Le sol de l'entreprise*, ed. ADEF, Paris, 199p.

ARU (1985) *Annales de la Recherche Urbaine*, no. 28.

Auguste-Thouard (1988) *Revue Urbanisme*, no. 228, December.

Auguste-Thouard (ed) (1990a) *Le marché de l'immobilier d'entreprise 1990, Tome 1, Le marché de l'Ile de France*. Auguste-Thouard, Paris, 69pp.

Auguste-Thouard (ed) (1990b) *Le marché de l'immobilier d'entreprise 1990, Tome 2, Le marché des métropoles régionales*, Auguste-Thouard, Paris, 159pp.

BIEN (1989) *L'immobilier d'entreprise dans l'agglomération grenobloise*. doc. pol., Grenoble, February, 28pp.

Bouchers, M. (1989) *Stratégies industrielles mondiales*, Paris: la documentation française, collection. *Les cahiers français* no. 243, Oct-Dec, 68pp.

Bouinot, J. (1987) Le défi de la crise économique et du décline urbaine, in *L'action économique des grandes villes en France et á l'étranger*, Economica, Paris, 316pp.

Bruhat, T. (1990) *Vingt technopoles, un premier bilan*. La documentation francaise, Paris, 188pp.

Chambre de Commerce et d'Industrie de Grenoble, Fondation pour la communication locale (1989) *Villes, Entreprises: construire ensemble*, éd Berger-Levrault, Paris, 121pp.

Choay, F. and Merlin, P. (1988) *Dictionnaire de l'Urbanisme et de l'aménagement*, , Paris, 723 pp.

Comité pour l'Aménagement et l'Expansion Economique de l'Isère (1988) *Immobilier d'entreprise: quel financement en Isère*, doc. mult., Grenoble, November 122pp.

DATAR RECLUS (1988) *Atlas de l'aménagement du territoire*, La documentation française, Paris, 368pp.

DATAR RECLUS (1989) *Les villes européennes*, La documentation française, Paris, 79pp.

DATAR (1990a) *Vingt technopoles, un premier bilan*, La documentation française, Paris, 214pp.

DATAR, (1990b) *Une nouvelle Etape pour l'Aménagement du Territoire*, La documentation française, Paris, 152pp.

Dollé, in Boucher, M. (1989).

Heurteux, C. (1990) *Les zones e'entreprise*, éd des PUF, Paris (Collection *Que sais-je?* no 2547) 128pp.

IAURIF (1986) *Bâtiments industriels locatifs en région Ile de France*, 126pp.

INSEE (1990a) *Tableaux de l'économie française TEF 1990*, 14th edn, ed INSEE, Paris, 188pp.

INSEE(1990b) *Annuaire retrospectif de la France*, ed INSEE, Paris.

Lacaze, J.-P. (dir) (1985), *Les grandes friches industrielles*, La documentation française, Paris (Ministre de l'Equipment, DATAR, rapport du groupe de travail interministériel), December, 149pp.

Leloup, G. (1986) A nouveaux besoins, nouveaux produits *L'observateur de i'immobilier*, no. 3.

Le Monde, n° spéciaux sur l'immobilier d'entreprise, 22 juin 1988, 3 novembre 1989, 24 octobre 1990, 6 juin 1990, 17–18 fevrier 1991.

La Gazette des communes, 19 octobre 1985.

Ministère de I'Economie et des Finances (1989) Notes Bleues, no. 416, January.

Pecquer, B. (1989) *Le développement local.* (ed. Syros) Alternatives économiques, 148pp.

Punter, John (1989) France, in *Planning Control in Western Europe* (ed. HWE Davies) Department of the Environment, HMSO, London, 149–252.

Sauvez, M. (1989) Les documents d'urbanisme sont-ils des obstacles aux implementations?, in *Le sol de l'entreprise*, ADEF, Paris.

SCIC, G31 (1988) Concepteur, promoteur, maître d'oeuvre *Urbanisme*, **228**, December.

Thouard, J. (1989a) *L'immobilier en 1990: le règle du jeu*, êd. Jean Thouard, Paris, 371pp.

Thouard, J. (1989b) *L'immobilier en France: L'état du marché*. éd. Jean Thouard, Paris, 4ème trimestre, 227pp.

6

ITALY

Alex Fubini, Alberto Landi and Rocco Curti

Little has been written about the Italian planning system, the property market and the role of local authorities, in particular of regional government. Furthermore, research into these themes has not been as comprehensive as that carried out in other countries. The aim of this chapter is to address the Italian situation and to supply readers with background information on the Italian legal and political framework.

The great inter-regional variability in Italy means that policies that have proved effective in one part of the country may not necessarily produce the same results elsewhere. Therefore, as well as outlining the interaction among the various public authorities, this chapter will also describe how each region has chosen to apply national policy in its specific context. Italy has never had a national policy on urban development, and specifically none on the planning of industrial sites, with one exception described below. However, national policies were and are implemented for housing, transport, services, economic and industrial growth.

Public policies on industrial location and real estate address two spheres:

1. the economic aspects (funding, tax, training schemes, job creation, labour policies, subsidized energy, social security and pension benefit payments);
2. planning and development of industrial estates, facilities and utilities.

The economic aspects comes under central government; planning and development come under local authorities, i.e. city and regional government, with one notable exception. This is the case of southern Italy, the Mezzogiorno, where the central government intervenes directly.

In the late 1950s, legislation was passed with the aim of favouring industrial development in the south. These laws generated a policy of industrial zoning which was subsequently translated into ASI (*area di sviluppo industriale*) and industrial development plans. The government also resorted to special action plans in the event of emergency or natural disaster, as was the case with earthquakes.

This chapter will address the following issues:

1. local and central government in Italy;
2. laws and technicalities in Italian planning;
3. physical planning in industrial locations, including procedures and regulations;
4. economic, financial (preferential loans, low-interest mortgages, funding and grants), and fiscal policies for industry;

5. industrial property as an economic good;
6. case study: the Lingotto (re-zoning, conversion, and renovation; private ownership and plans to generate a new market value for commercial and industrial premises no longer in use, in an over-supplied market).

6.1 LOCAL AND CENTRAL GOVERNMENT IN ITALY

Planning permission in Italy is granted by the local authority, which is why a brief overview of the Italian local authority system may be useful. Unlike some other European states, such as Germany, Italy is not a federal republic: it has a centralized rule, counterbalanced by spheres of separate powers, known as autonomies. This means that the country has a two-tier power system with a local and a central level. The central or national government makes the laws and drafts national policies, as well as setting guidelines for local policies. Before 1970 there was a one-tier local government, the municipal governments or *comuni* – also called the administrative level – with the central government known as the political level. In 1970, consistent with the guidelines of the Italian Constitution, some of the central powers were devolved and regional governments were established as an important tool in law-making and strategic planning.

Five regions had already been created just after World War II: the special statute regions (*regioni a statuto speciale*, RSS). They enjoyed a high degree of autonomy and were either border regions or major islands: the Aosta Valley, Friuli Venetia-Julia, Trentino Alto Adige (the South Tyrol), Sardinia and Sicily. In 1970 the central government delegated or transferred some of its powers to the regions, which can consequently now legislate on some issues, one of the most important being planning and development. Delegated powers comprise the power to legislate, while transferred powers refer to the implementation of central policies, such as health, education and energy.

The newly created regions, known as ordinary statute regions (*regioni a statuto ordinario*, RSO) were empowered to plan, make by-laws, statutes and regulations for the bodies and institutions under their authority; set local authority boundaries; decide on the regulations for the traffic police; draft the Forestry and Rural Police Regulations, the Fairs and Market Rules and By-laws. They are in charge of charities and assistance; vocational training; local authority museums and libraries; tourism; hotels and catering; regional public transport; lakes and inland water navigation; berthing, harbours and ports on inland waters; hunting, and fishing in inland waters; quarries; agriculture and forestry; crafts. Statutes give regions both economic and territorial planning powers. In the decade since they were first established, one of the main tasks of the RSOs has been to draft regional development plans (RDPs), very much like the UK's structure plans, following regional planning regulations consistent with national guidelines laid down by national government. However, since the early 1980s, planning has played a less important role since government funding has been more for general administration, with resources fragmented and scattered.

Until the 1973 tax reform, councils had strong levying power, and local taxation accounted for 60% of their income. Currently the percentage has dropped to just over 10% and local authority spending has become dependent on central government transfers. The amounts are calculated according to the size, population and socio-economic classification of each local authority. Town councils in Southern Italy are granted special conditions. Regions receive about 7% of GNP, provinces 4% and municipalities 0.5%.

Regions are not in charge of industrial policies, since central government is in charge of the direct and indirect promotion of economic activities. Effectively, RSOs are left with territorial or regional planning, specifically with the possibility for creation of industrial estates or centres to attract business. Several regions have created or relocated industrial estates, especially in the north of the country, between 1975 and 1980. Initially policies were aimed at creating a small number of large, well-served and connected industrial areas in the outskirts or in marginal areas to counteract the over-concentration in urban or metropolitan areas.

However, from the early 1980s, the failure of these policies led the authorities to opt for smaller and more evenly scattered centres. Currently regions are able to offer a range of industrial and commercial premises, and several RSOs have set up public corporations to develop and manage these industrial areas and estates. These development corporations work jointly with the municipality of their areas, associations of industrialists and with business consortia. This is how a region can use the only effective tool it has in industrial policy-making to promote well-served industrial and business areas and to make them available at reasonable prices. Unlike local authorities, the central government has, since the inter-war period, pursued industrial policies aimed at developing the country's economy – and manufacturing industry in particular. The government has done so through funding, endowment financing, easy credit or tax breaks, by reducing the employers' share of social security and pension benefits, by giving research and development grants, and in some sectors by creating trade barriers.

All these policies state their commitment to planning, but the practical outcome has been a greater degree of fragmentation, scattering of resources and an ongoing negotiation-mediation process. The causes can be traced back to the complex decision-making process, to the inconsistencies and contradictions in the many laws and to the fragmentation of the political system. Municipal governments or *communi*, unlike regions, cannot legislate although they enjoy a high degree of autonomy in implementing national and regional laws within their boundaries (traffic regulations, schools, developing and building, hospitals, roads, etc.). Local government also includes provinces initially created following the Napoleonic model, as the executive branch of the central ministries. Although they were conceived as merely technical-bureaucratic institutions, following the reorganization of local government in 1990, they have now also acquired some planning powers. The country is divided into 21 regions, 95 provinces and about 8800 municipalities. The 1990 law which reorganized local authorities does not affect planning regulations and their main procedures.

Regions, provinces, and town councils (municipalities, or *communi*) are elected bodies. The region has an elected council which legislates and fulfils the duties set out by central government and basically acts as a local parliament. The regional government itself is the executive; the President is the legal representative and signs all the laws. The province is administered by a council which elects a government and a president. Provinces are responsible for the building and maintenance of provincial roads; for public works (water works, geological work, ports and infrastructures); and for mental health. The provincial authorities can also form consortia or *ad hoc* bodies to manage services such as health and vocational training. Provinces have fragmented and sectorial responsibilities compared to municipalities and regions: their function will change with the implementation of the 1990 Reform of Local Authority Act (Law 142), whereby provinces become part of a network of local authorities (*autonomie locali*). They will cease to act as a peripheral arm of the State and will become an authority in between the other two (region and municipality) with a horizontal (sectorial) responsibility such as environmental defence of the soil, habitat, energy, monuments and cultural heritage, and parks.

The municipality has an elected council which in turn nominates the city government and the mayor. The town council or municipality is the basic unit of government with responsibilities in the administration.

Some of these duties will have been delegated by the central state and region; others are specifically local. Such responsibilities may be either optional or compulsory and they include town planning and building permits as well as the municipal police, public health, refuse collection, social services, and basic urbanization (street lighting, sewers, water, etc.). The town council has a mayor (*sindaco*) who head a government (*giunta*) and chairs the council (*consiglio*).

Local government was reorganized in 1990, following Law 142: municipalities now draft their own statutes or charters. The law also places new powers and responsibilities in the hands of local government, and creates a new authority, the new metropolitan government for the nine main Italian cities and their metropolitan areas. These metropolitan governments will also be responsible for planning and management of the area.

6.2 LAWS AND TECHNICALITIES IN ITALIAN PLANNING

At national level, there is a series of parliamentary or statute laws setting the legal framework, defining both the nature and the hierarchies of planning tools as well as outlining procedures. Regions have been acting within this framework since 1972, developing planning instruments and setting guidelines for their area. According to Italian planning law, planning activity can be implemented at the following three levels: region; commune; sub-commune. These levels correspond to different forms of plan: territorial plans (*piani territoriali*, PT); general development plans (*piani regolatori generali comunali*, PRGC); and local action plans (*piani attuativi*). These plans are differentiated not only by spatial level but also by the role they play in the context of land-use planning.

Territorial plans (PT) define the objectives and programme of strategic guidelines for a large area, usually a region or large part of a region. All territorial plans are general structure plans and in principle they do not establish zoning. They set out aims, policies and recommendations as well as development guidelines and constraints, defining protected areas and land reserves for infrastructure, roads etc. They comprise: (a) global and comprehensive structure plans; (b) sector-specific plans for parks, landscape, environment, transport, etc. Such plans are drawn up by regional governments. In some cases they are approved by committees, while in others, including Tuscany for example, they require approval by the full regional council.

General development plans (PRGC), covering the entire area of a municipality, define the land use of the commune in accordance with the guidelines of the territorial plan. These PRGC plans are almost always municipal, but are in some cases drawn up for several adjoining communes and are then known as *piani regolatori intercomunali*. They consist of a zoning map and written statement, which provides a detailed basis for implementation and control of development. The PRGC is the key planning instrument available to the commune administration. Through it a commune can set its objectives and guidelines for development and conservation. The main features of the contents of a PRGC are:

1. communication networks and infrastructure;
2. zone subdivision of land;
3. land reserved for public use, state buildings, public buildings and development for social purposes;
4. development ratios;
5. restrictions on development for reasons of environmental protection, historic preservation etc;
6. plan implementation norms (administrative and technical procedures for the implementation of the PRGC, so-called *norme tecniche di attuazione*, NTA).

Local action plans (*piani attuativi*) do not extend over the entire municipality. They are detailed plans referring to specific areas or zones. They usually cover small areas defining physical change in accordance with the general plan. There are several kinds of local action plan, depending on the sector of building concerned (private residential, public housing, industry for example). Some of the possible categories are: *piano particolareggiato* (PP), a very general type of local plan; *piano di zona* (PdZ), a local plan for low-cost and public housing; *piano esecutivo convenzionato* (PEC), a plan for land subdivision into residential or productive plots, drafted directly by private land and property owners; *piano di recupero* (PdR), an urban renewal plan; and *piani per insediamenti produttivi* (PIP), a plan for productive activities. Some are drafted by the local authorities (PP, PdZ, PIP) and some may be drafted and presented by the private sector (PEC).

Procedures for local (general and action) plans are as follows. Plans are drafted and adopted by the city government, and then either approved or rejected by the region. In some cases, and in some regions, action plans can be approved directly by the municipality without reference to the region. Under planning legislation, the planning process at the commune level has two phases: (a) general planning through the PRGC which sets overall objectives but no time schedule; (b) detailed planning, through the action plans, putting into effect the objectives of the PRGC. In Italy, only a small number of detailed plans have been prepared, approved and implemented.

Recently another implementation tool has been introduced by national legislation, in order to programme the implementation of proposed development: the *programma pluriennale di attuazione* (PPA) or multi-year implementation programme. The law requires all major communes to draw up a PPA as an implementation tool for general development plans. The PPA cannot propose new planning policies; rather, it is meant to provide an indication of priorities by introducing the time factor into a general plan where there is no time limit for proposed changes. The PPA has three main purposes:

1. to coordinate proposed changes by stating a time-frame for their implementation;
2. to link urban planning at the city level with the economic and financial planning of the commune;
3. to enforce the implementation of the plan by compulsory assembly of the areas which should be developed according to the plans themselves.

In practice the PPA has been little more than a shallow list of 'intentions to build' as declared by property owners because of the very weak enforcement power of local authorities due to their lack of financial resources. Nevertheless, major developments cannot be implemented if they are not included on the PPA list. The PPA is periodically drafted by the commune every three to five years, and can be reviewed each year when the commune declares its annual budget.

6.3 PHYSICAL PLANNING IN INDUSTRIAL LOCATIONS

In Italy, building or planning permission is required to introduce any change in land or buildings. Permits have to be consistent with general plan zoning and, as the case may be, with detailed plan specifications too. If permission is granted, the permit fees have to be paid in proportion to the floorspace of the premises and according to location. Regional laws state how and where the fees have to be paid, while city governments establish the amount, which depends on the zone within their administrative boundaries. These fees are known as *oneri di urbanizzazione*. The sum to be paid is calculated by the city council following a regional statement which defines criteria and maximum and minimum thresholds. Fees can vary, depending on the size and characteristics of the development. For commercial and industrial premises, the fees also depend on the goods and merchandise manufactured and, for a major industrial development, local authority and industry can negotiate the fee.

An entrepreneur, or anyone wishing to build industrial premises, has to comply with the following procedure. A plan has to be submitted to the local municipality. It will be examined by a municipal committee (*commissione igienico-edilizia*) which will express its opinion on all the plans which entail changes in the use class. The committee has to express its opinion though the mayor is not bound by it. The mayor has the authority either to grant or reject plans and has to sign all building permits. If granted, the permit entails the above-mentioned tax or charge relating to the size of the building (usually according to the floorspace), the kind of building and its location and zoning.

Several other bodies (Provincial Health Authority, the Fire Brigade, the Geological Service in earthquake areas and the environmental authorities for pollution risks) are consulted before granting any authorization. Since the Merli Act (Law 319 of 10 May 1976, the first and most important law regulating water resource management and pollution control), water, air and, more recently, also sound, anti-pollution regulations have to be adhered to carefully.

In the event of permits for 'large building sites' (as defined by the law) the environmental impact has to be assessed in advance. Since the Prime Minister's Decree (No. 377 dated 10 August 1988), Italy now complies with EC directive EEC 85/337 requiring the environmental assessment of certain private and public works which may have an especially strong environmental impact.

The following are tested for environmental impact, in addition to infrastructure which is usually a public works development:

1. oil and natural gas refineries;
2. thermal power plants and thermal production plants when power exceeds 300 MW;
3. storage and processing plants for radioactive materials;
4. steel mills;
5. asbestos production;
6. chemical plants;
7. refuse processing plants.

A subsequent decree (dated 27 December 1988) introduced the technical regulations and standards used in assessing these kinds of project. These standards were used to define: (a) the content of an impact study, the criteria used in assessing compatibility, and the papers to be supplied;

(b) environmental factors and scope; and (c) the criteria to be used in drafting specific studies for each category.

The municipality is the only authority that can grant planning permits, and decisions must be consistent with the town plan. We shall now analyse some examples.

If the building requires a permit according to the town plan, the permit can be obtained as long as the building conforms to the zoning and to the development plan for the site (*norme di attuazione relativa all' area specifica*).

The site may be listed as one requiring an action plan (*piano esecutivo*), in which case there are two possibilities. First, the action plan may be drafted by a private entrepreneur or developer covering a wider area than the specific site for which permission is being sought. If an action plan has been drafted, the procedure is as described in the previous paragraph; if not, an action plan is required and the various owners will have to submit one as they are the only ones entitled to do so (*aventi titolo*). In the event of more than one owner, they will have to agree to a joint action plan. This action plan must then be approved by the municipality and, according to local regulations, it may have to be approved by the region too. Once approved, each owner – if there is more than one – can apply for a building permit as described above.

There is also the case of a public action plan, that is to say a plan drafted directly by the municipality – which in some areas may have to be approved by the region too.

There are several kinds of action plans (*piani esecutivi di iniziativa pubblica*). The ones for industrial sites are known as *piani per gli insediamenti produttivi* (PIP) if the local authority has previously bought the site, and the sites – either purchased directly or compulsory purchased – are initially divided up into sites or plots and then urbanization work is carried out, that is the area is supplied with the necessary infrastructures, roads, connections and utilities industry may require. The sites are subsequently allotted. A portion of the area, not less than 30%, is sold leasehold with a lease not exceeding 99 years to be renewed for another 99 on request. The remaining property can be sold freehold. In both cases there are charges and transfer taxes. Municipal authorities can set prices for both leases and freeholds for the original site.

Development and management consortia or corporations are often created and act according to agreements signed with the local authority. The latter may choose to delegate all the implementation phase to the consortia. In many cases the creation of these PIPs is the result of the regional indirect incentives mentioned above, within the scope of their authority. In the event, the municipality receives the funds from the region to set up the industrial estate (i.e. purchase the land and initiate infrastructure). Regions operate through *ad hoc* public financial institutions which in turn may choose to act by creating companies – mostly publicly owned or with a local authority majority stake – jointly with credit institutions, banks and developers themselves. The aim of these companies or corporations is to create a serviced industrial estate and to market the sites.

On the whole, the availability of industrial sites in northern Italy is below actual demand, while the southern parts of the country still offer interesting opportunities especially thanks to the special government funding of development and aid agencies operating in the south. However, northern cities have a number of former industrial sites (in very central areas too) currently not in use, as is the case of the Bicocca area in Milan, or the Fiat Lingotto in Turin (see the case study in section 6.5). Incentives for BICs (business innovation centres) have not proved so successful in Italy and very few have taken off. There are exceptions in Tuscany (the Prato district) and in a very limited number of regions mostly in northern Italy, European legislation and incentives for BICs apply to Italy as well.

6.4 INDUSTRIES, MARKETS AND LOCATION

Italy has a population of 57 million inhabiting a surface area of 301 223 km^2 , which makes it one of the most densely populated countries of the EC (189/km^2). Furthermore, 67% of its population live in urban areas and the country is connected by a strong urban network with several large metropolitan areas and cities whose population exceeds 1 million, i.e. Rome, Milan, Turin and Naples. Most of Italy is covered by mountains (the Alps and the Appenines), so that only about 30% of the country's total surface can be worked as agricultural land. The presence of mountain ranges also affects the transport system.

Italy's economic structure has undergone great economic changes over the past decade, both in terms of product and of employment patterns: the tertiary sector is now larger than the manufacturing one, and the country's overall GNP ranks sixth in the world. Car and metal works, along with the chemical industry, are the single largest employers in industry, followed by publishing, clothing, electronics and rubber (Table 6.1) There is still a large gap between the south and the north of the country (Table 6.2), and it is not likely to disappear in the short term. The divide affects several facets of social and economic development:

local economic systems, income per head, lifestyles, consumption levels, infrastructure and services.

Table 6.1 GNP and employment data by sector

Sector	GNP ($US m)	Percentage of total	Employed	Percentage of total
Agriculture	30 608	4%	2 229 452	10.7
Industry	221 908	29%	8 813 372	32.7
Building	38 260	5%	–	–
Others	474 424	62%	11 793 176	56.8
Total for Italy	765 200	100%	20 836 000	100

(Source: ISTAT, 1988)

Table 6.2 GNP per capita: EC index by region, 1988 (Europe = 100)

Region	Area	Index
Aosta Valley	North-west	156.7
Liguria	North-west	133.9
Lombardy	North-west	133.5
Emilia Romagna	North-east	130.7
Piedmont	Northwest	125.5
Friuli V.-Julia	North-east	117.3
Tuscany	Centre	116.4
Trentino A.-A.	North-east	115.2
Latium	Centre	110.1
The Veneto	North-east	106.2
Umbria	Centre	102.7
The Marches	Centre	100.4
The Abruzzi	Centre	84.9
Molise	Centre	79.1
Sardinia	South	77.1
Basilicata	South	76.0
Apulia	South	74.5
Campania	South	72.7
Calabria	South	62.3

(Source: EC and Europa (1990) *Il Sole 24 Ore*, Milano 1990)

The Po Valley (*Pianura Padana*) is also known as the Pomegalopolis: it has the highest concentration of manufacturing, services and infrastructure. In fact the plain houses the majority of the country's industry and tertiary sector, with a strong international dimension of northern Italy as well. The Po Valley lies along the West-East axis; it crosses the regions of Piedmont, Lombardy, Venetia and Emilia-Romagna.

6.4.1 The building industry and its market

Compared to the locations of Italy's industry only a few years ago, things have changed a lot: new manufacturing areas have sprung up alongside previously existing industries in the wake of incentives granted to some districts and thanks to the optimization of entrepreneurial potential. They are to be found in well-served and well-connected areas, normally outside metropolitan areas. Sixty per cent of Italy's 1 005 408 productive units are in the north of the country; 100 of the 142 enterprises with more than 1000 employees are also in the north compared to 40 in the centre and south (Table 6.3).

Table 6.3 Industrial plants: data by region, 1981

Region	Area	Number of plants	Percentage of total
Piedmont	North-west	97 481	9.70
Aosta Valley	North-west	2 402	0.24
Lombardy	North-west	207 794	20.37
Trentino A.-A.	North-east	18 129	1.80
The Veneto	North-east	108 281	10.87
Fruili V.-J	North-east	27 839	2.77
Liguria	North-west	25 802	2.57
Emilia Romagna	North-east	112 523	11.18
Tuscany	Centre	98 547	9.80
Umbria	Centre	16 270	1.62
The Marches	Centre	42 008	4.18
Latium	Centre	47 985	4.77
The Abruzzi	Centre	18 815	1.87
Molise	Centre	5 460	0.54
Campania	South	41 885	4.18
Apulia	South	45 163	4.48
Basilicata	South	8 833	0.88
Calabria	South	16 441	1.64
Sicily	South	45 734	4.55
Sardinia	South	19 807	1.97

(Source: ISTAT, 1988)

One of the characteristic features of this pattern of decentralized location of industry is that growth occurs around one main industry – as for instance is the case of Prato and the textile industry, or of electronics in the Tiber Valley, and of knitwear and clothing in the Carpi district (*Emilia Romagna*). Small and medium-sized enterprises have proved able to adjust their company strategies to changing needs, shifting resources and space. Hence in some areas there is a shortage of manufacturing plants and storage space, while in other districts there is an over-supply and prices are much lower. Industrial building and developments depend on the growth and expectations of local economies. Local property market trends are also strongly influenced by competition in its sector. The pattern is more even in the outskirts of metropolitan areas, where business and manufacturing sectors are more highly specialized and there is a more diversified picture. The result of the recent economic growth is an increase in commercial and industrial building, especially in Lombardy and the Veneto: the two regions accounted for 40% of all new industrial buildings in 1987.

This uneven market has produced a strong variation of purchase prices for an average size industrial building range from LIT 200 000/m^2 – for which you would get an iron frame in the outskirts of a declining industrial area – to LIT 1 000 000/m^2, which would buy a pre-fabricated or reinforced concrete building in a thriving area. Yearly rents per square metre range from LIT 20 000 to 70 000. Office space is roughly three times the price of production plant costs.

Table 6.4 New industrial premises: building data by area 1987

Area	Number of buildings	Building volume (m^3)	Average (m^3)	Percentage of total
North-west	1 922	14 808 288	7 705	33.07
North-west	2 017	14 206 209	7 043	31.72
Centre	1 356	8 279 371	6 106	18.49
South	1 306	7 490 125	5 735	16.72
Total (Italy)	6 801	44 784 074	6 784	100.0

(ISTAT, 1988)

Table 6.5 New industrial premises 1985–1987

Year	Volume (m^3)	Growth
1985	35 074 280	–
1986	32 917 158	−0.15
1987	44 784 074	+36.05
1988	47 591 712	+6.27

(Source: CENSIS Servizi, 1988)

6.4.2 Commercial and industrial property

Market actors

Actors on the commercial and industrial property market vary according to the size of the premises, rather than the size of industrial estate as a whole. The future owners or users usually manage the building or sale of the properties in the case of medium-sized or large plants. Large premises are exclusively commissioned or developed by their intending users. Properties of this kind are either advertised or can be found in the main national dailies, and above all in the financial daily *Il Sole 24 Ore*. Regional financial institutions provide another channel, since they both assist firms relocating to industrial premises no longer in use and promote serviced areas. Where small to medium-sized premises are concerned, local builders will act as speculative developers, both building the structure and selling the units.

Agents dealing with commercial and industrial premises control about 30% of all the market and need to be contracted, especially if existing premises or old buildings are sought. Legislation has recently been passed, regulating real estate agents: there are now 13 000, all small businesses except for the 10 that operate nationally.

Purchasing real estate in Italy

According to the law of the land, in Italy property can be owned by the state, the regions, provinces, municipalities, public bodies or private citizens. By virtue of its nature or use property is either non-transferable (inalienable) such as some common land – coastlines for instance – or transferable: property can be granted to private citizens, although such permits or concessions can be withdrawn at any time. Furthermore, a municipality can grant a lease (*diritto di superficie*) rather like the British leasehold, but generally limited to publicly owned land.

Land is divided into units and registered in the Land Cadastre or Registry (Nuovo Catasto Edilizio Urbano, NCEU), which is held in each province. Cadastres list all the deeds, the boundaries, the size and description of all real estate property. Cadastres also register minimum values of properties for tax purposes. Ownership rights can be either on a new building, or can be obtained by transferring property deeds (*titoli di proprieta*) from the previous owner. Ownership can be obtained by signing a contract, or by inheriting it from the previous owner. The state or its authorities or bodies may issue compulsory purchase orders in the event of the property being of 'public interest', or in lieu of debts or moneys owed.

The purchase contract is an official document which must be drawn up by a notary (*notaio* – a public officer who authenticates deeds and also performs many of the duties a solicitor might carry out) who will then register the sale and purchase. Copies of contracts are kept at the Registry of Real Estate (*Conservatorie dei Registri Immobiliari*). These contracts specify the rights and limits attached to the deeds: the Registry can be consulted for reference if the purchaser wishes to view previous sale acts and contracts.

Building commercial and industrial premises

The ownership of a plot of land or of a building does not *per se* entail building or conversion rights, or rights to change the volume, surface, use or outer facades. Planning permission (*concessione edilizia*) is granted to the owner who has to submit the plan to the municipality where the site and, according to the zoning, the regulations and guidelines are set out by the local authorities. Leaseholders of public soil may also apply for planning permission. It is therefore essential to be clear about the use one wants to make of a building, and about whether it is compatible with existing regulations and zoning before purchasing it. Furthermore, the environmental regulations, refuse disposal, energy consumption and fire regulations also have to be carefully considered and abided by.

Some municipalities have adopted the so-called PIPs – industrial action plans – where services and infrastructures are supplied and the industrial estate is promoted and managed by the local authority.

Prices per square metre range from LIT 20 000 to 50 000 plus urbanisation charges (LIT 20 000 – 10 000/m^2) and building costs (about LIT 400 000/m^2 for a prefabricated building).

Renting industrial premises

In Italy renting industrial premises, as opposed to buying them, is a viable alternative. The market caters for small and medium-sized premises and accounts for roughly 50% of all industrial premises advertised in national dailies.

There is a Rent Control Act (*Legge sul Equo Canone*, 392/75) which governs the renting of residential, commercial and industrial premises. In the case of residential buildings the six-year contracts are to be automatically renewed or extended for another six years unless otherwise stated; the rent itself is agreed upon when the contract is signed or renewed. The rent may increase only by the inflation rate for the duration of the contract. The rent may be increased by a maximum of 75% of the inflation rate. Inflation is calculated according to the rate published by the Italian Bureau of Statistics (ISTAT) once a year.

Company and tax law

Italian company law recognizes two kinds of taxable company: partnerships (*societa di persone*) and incorporated companies (*societa di capitale*). In partnerships, partners are personally liable. Incorporated companies (limited, shareholding and joint-stock) can have capital stock from individual members and from other companies; there is no direct liability, linking the commitments and undertakings of the company and stock or shareholders' personal estates, since their liability is limited to their stake. The following rules apply exclusively to incorporated companies:

1. direct taxation of company profits (IRPEG, taxation on legal entities or legal persons, and ILOR local income tax) at about 46.4% of the net profit;
2. indirect taxation such as VAT, company registration tax, business tax, and tax on the premises (a sort of rates).

In calculating profits for direct-taxation purposes, depreciation includes real estate depreciation up to a maximum of 3% of the value of the land and property with total depreciation taking 34 years. In the event of a sale there is a double tax: appreciation (the difference between the purchase and sale values minus residual depreciation) is considered as income (ILOR and IRPEG); there is also a type of capital gains tax called INVIM.

INVIM is a tax on the presumed increased value of the property. Companies and corporations pay INVIM either when they have owned the building for 10 years (INVIM *decennale* or 10-year INVIM), or when the property is sold. The taxation is calculated as follows. The purchase price and some of the important costs (if any) are summed and then deducted from the sale price. The tax is a percentage of the result: the greater the sum, the higher the percentage. In the event of 10-year INVIMs having been paid, the final tax is calculated at the sale on the basis of the last INVIM to be paid.

6.4.3 Industrial locations and funding for enterprises and business

Industrial locations: policies

One of the main objectives of the current economic policy of the Italian government is to redress the imbalance between poor and rich areas in the country. The industrial development of the poorer areas is one of the main targets. The development zones are usually identified with what is called the South, that is the regions south of Rome, the two largest islands (Sicily and Sardinia) and other depressed areas (earthquake zones, mountain areas, etc.). One of the main incentives consists of offering industry lower running costs: lower-priced energy, lower cost of labour, easy loans and capital account funding. The co-ordinating agency is called AGENSUD, and regional governments may add incentives for development. These will be handled by regional financial institutions, development agencies with an aim to promote local businesses and even hold shares in them. Some regions also include the development and funding of AIA (*aree industriali attrezzate*) and relocation projects.

Real estate financing

Mortgages and what in Italy are known as 'real estate leasing contracts' are the tools most frequently used to finance the purchase or building of real estate. In the case of a mortgage, the borrower obtains funds to purchase a property; mortgages are mostly 10-year loans, covering no more than 70–80% of the value of the property. A mortgage on a property is a guarantee for the lending agency and is cancelled when the entire sum is paid back. In the case of properties built on leased land (*diritto di superficie*), banks do not readily grant these mortgages, since they favour mortgages on freehold properties since the real estate is the guarantee. In the event of payment default, the credit institution can reclaim the property and gain full ownership rights over it. Interest rates on mortgages are currently about 14% (7.5% after inflation is taken into account), although easy or subsidized mortgages can be as low as 8.5% (2.3% in real terms); alternatively they may be partly funded as capital account. For taxation purposes, the interest on the loan or mortgage is considered a cost of the company or business, following the criteria described earlier on depreciation.

In Italy the term 'leasing' corresponds to a particular financial tool, a rather typical type of funding: although the English word is used it must not be confused with English lease and freehold. A 'leasing' company purchases or builds premises (or any other merchandise) on behalf of the customer and according to the contract they have signed. The property or good is then given to the customer to use; at a given date, specified in the contract, the property or good will be offered to the said customer at a pre-established price (*prezzo di riscatto*), and the customer may purchase it, paying in instalments. Instalments are determined in relation to the financial depreciation of a loan corresponding to the value of the property or merchandise. Interest rates range between 15 and 20% and easy terms or grants are possible.

The redemption price (*prezzo di rescatto*) ranges between 1% and 10% of the initial value of the property or good. The redemption price becomes the book value. According to current legislation a 'leasing' contract on a property must last at least eight years. The advantage of this system compared to a traditional mortgage is that since it is offered as a service (the property is initially rented, and only subsequently sold), instalments can be written off as costs and are thus deductible from taxable income. In the case of mortgages only a part of the interest is tax-deductible, and consequently in spite of 'leasing' rates being higher than mortgage ones, tax breaks are more effective for the former, making it financially a more attractive proposition.

6.5 CASE STUDY

This case study of the Lingotto illustrates re-zoning, conversion, and renovation: private ownership and plans to generate a new market value of commercial and industrial premises no longer in use, in an over-supplied market.

The Italian industrial property market consists of two equally important halves: the building of new industrial and commercial premises and the renovation or conversion of old ones, especially in the main urban areas, where many former plants and industrial estates are now no longer in use. The late 1970s and 1980s witnessed a great change especially in northern Italian centres such as Turin, Milan and Genoa. New developments in finance and production released large areas in the centre of these large industrial cities, giving rise to a new industrial property market. Turin and Milan alone had 6 000 000 m^2 of industrial premises no longer in use, a vast amount compared to the size of the cities and the increasing shortage of space in urban areas. The private sector can develop these premises, and the local administration can make use of them too, with the aim of upgrading urban areas. These changes are favoured by the present trend in urban planning, based on fewer constraints, and a more pragmatic outlook. Urban planning lays down the guidelines, which are then implemented by *ad hoc* plans – very much like 'action plans' – exclusively applied to areas where conversion is a viable proposition. Such plans are finalized through a negotiation process involving local authorities and developers.

The property market has thus shifted from a notion of inflexible values – not to be changed once the master plan and its rules had been established – to a more flexible system with values that can be negotiated in relation to public and private needs and requirements. The following are the aspects most deeply affected by the recent change in the system:

1. the economic and legal framework regulating the relationship between local authorities and private sector which are currently ruled by the compulsory purchase order laws (*espropri di pubblica utilita*);
2. the nature of the various agreements, such as concessions and licensing, and the creation of public-private companies;
3. the outcome of conversions, which increasingly appear to be the result of negotiations between public and private sectors, where services and environmental standards are exchanged, use class and building indices. In other words the buildings received a more profitable use class classification and were granted higher indices in exchange for other areas and/or for the development of social services and/or buildings of public use.

Turin is one of the most important towns in northern Italy; *commune* population about 1 million; greater Turin 2.2 million; it is a highly industrialized area, the centre of the car industry. Turin illustrated the above-mentioned points, since there are several industrial areas and estates no longer in use, many of which are quite central. It is essential to foreign investors to understand the mechanisms underlying development and conversion laws and regulations. This case study deals with one such example, where one of the oldest and most important industrial plants of the 1920s can be found: the Lingotto Fiat Plant.

6.5.1 The Lingotto

The Fiat Lingotto plant was built between 1911 and 1920. It covers a surface area of 18 ha, with a total volume of 1.7 m^2. It is very close to the city centre, along the main railway line. The building is a very good example of the then *avant-garde* architecture, and was modelled on the Detroit Fort plant; it is thus an interesting example of industrial archaeology. The plant was closed in 1982 and since then Fiat has tried to create a favourable political and administrative situation to maximize the value of the plant.

Fiat managed to swing public opinion in favour of the conversion of the Lingotto plant, shifting the debate away from the property market and employment aspects, and stressing the positive benefit the new use of the plant would have on the city as a whole. In 1982, Fiat organized an international competition of ideas and plans for conversion. In 1984, an exhibition of 20 of these plans for the Lingotto plant (20 *progetti per il futuro del Lingotto*) was held, and the company got permission to use Lingotto as a venue for the following year's Motor Show (*Salone dell'Auto*). They may have favoured the listing of the central building by the Ministry of Cultural Heritage and Environment (*Ministero dei Beni Culturali e dell'Ambiente*); the conservation order stresses its architectural and historical value, and means the plant is listed as a building of public interest. In 1985 the town council included the Lingotto amongst the areas of strategic importance in the new Master Plan.

6.5.2 The market value of the industrial plant

Fiat suggested Lingotto might be converted into a building for 'public use', the first step being the creation of a mixed, ie private-public company. Fiat's suggestion was that its capital contribution should include the market price of the premises. This meant that the first, though not necessarily the most important item on the municipality-Fiat negotiating agenda was the market value itself. One of the hurdles concerned the criteria to be used in establishing the market value of the Lingotto plant, considering the value of the building itself was very low, and that there were too many unanswered questions to assess the conversion value.

Local authority officials assessed the value according to the law on industrial and commercial property, in other words by following the compulsory purchase order procedure (i.e. the value of the land according to agricultural land rates, plus the value of the buildings themselves, urbanization costs). Thirty per cent was added to the price as compensation in lieu of a voluntary transfer, and 23% as relocation costs, bringing the total up to quite a good price (about LIT 250 000/m^2). According to these agreements the overall price of the Lingotto was to be LIT 45bn. However, when the details of the mixed company with a capital of LIT 45bn were finalized, Fiat and the municipality agreed on a lower price, LIT 23 625 m.

When calculating the market value, the parties decided to adopt the system commonly used for industrial plant relocation. In the event there was no actual relocation but simply a closing-down of that production plant. However, relocation criteria were adopted just the same.

The discount, if that is what it was, can be explained by the fact that Fiat was very interested in changing both the use class, from industrial premises, and the zoning; in fact if the Lingotto plant had been left as an industrial building it would have had no 'market value' due to the over-supply of industrial and commercial properties in the city. Clearly the market value of the Lingotto plant would depend on the re-zoning of the site and on the plot ratio or building index (the built-up volume for a given area): these were the real issues at stake in the negotiation process, and the ones which would determine the outcome of the whole operation.

Fiat also managed to envisage possible conversions and new uses, as was revealed by the feasibility study the local authorities required before drafting the detailed plan. The plan ended by re-tracing the Fiat 'decision', making the various steps official including the re-zoning. In 1988 the new detailed plan was approved: six years had elapsed since the announcement of the plant closure, and three years had passed since the inclusion of the area amongst the strategically important zones. This is a relatively long time, unless one takes into account the complexity due to the uncertainty of the political situation and to the change in attitudes towards planning.

6.5.3 Zoning and standards

The influence of the Turin car industry extended beyond the local level: the company carried its influence with the Ministry of Cultural Heritage which changed the constraints of the listing, granting more flexibility. As mentioned above, re-zoning and use classes in the detailed plans at the new planning standards were the real issues at stake. Negotiations were based on the feasibility study commissioned by the municipality of Turin, which in spite of being a local authority study turned out to be a purely financial analysis not taking into account the effects it would have on the city. The outcome of the entire operation effectively depends on the plan, which selects the activities for the area and introduces a high degree of flexibility. Thanks to the plan's modular approach, action may respond to needs and evolve according to the situation.

It was a two-tier approach by functions: there were so-called 'strong' functions, yielding a higher profit and which were especially important to Fiat, that is the exhibition centre acting as a pivot for the entire project. The weaker functions, with a lower level of profitability, included the University and the incubator – a structure aimed at budding economic and financial businesses. The identification of strong and weak functions, and of dividing activities into private and public uses, is justified by the size of the available space and by the difficulty in foreseeing how many private businesses would actually be attracted. According to the plan, 188 900 m^2 would be for office, managerial and executive space (65 000 m² for the exhibition centre, 9500 m² for a conference centre, 37 000 m² for offices and 9000 m² for retail businesses; 59 000 m² to the university and training; 2200 m² to cultural activities; 7200 m² for other offices in the Palazzina (the original plant); another 17 000 m² for production and manufacturing; and 15 200 m² for residential space. The space for public services totalled about 204 000 m², that is to say according to the law and to planning regulations; this too was negotiated at length.

6.5.4 Re-zoning and conversion: economic outcome

Feasibility studies were carried out according to strictly financial criteria, using a model able to test different hypotheses for each function, considering the type of contract, rent or sale, the forecast sales and tenancy costs, at the rates of the time, considering the yearly depreciation of the premises and also in relation to the yearly increase in rents.

Table 6.6 Building costs estimate and market values

Use	Building costs *a*	Market value *b*		Percentage profit *c* = 100 (*b*−*a*)/*a*	
		Min	Max	Min	Max
Exhibition centre	62 454	74 945	74 945	20	20
Conference centre	14 229	17 075	17 075	20	20
Innovation and discovery centre	40 407	49 950	59.940	23.6	48.3
Incubator	13 470	28 840	24 840	84.4	84.4
University	61 361	69 966	80 361	14.0	31.5
Services and business	9 379	13 770	17 010	46.8	81.3
Residential	17 781	20 520	24 624	15.4	38.5
Car parks	53 890	50 867	50 867	−5.6	−5.6
Total	272 971	321 933	350 031	17.9	28.2
Cultural activities	2 828				
Tower restaurant	1 395				
Helicopter landing pad	2 862				
Grand total	280 056				

(Source: Turin Municipality feasibility study. NB: The feasibility study did not indicate market values for the cultural activities, the tower restaurant and the helicopter landing pad.)

In the event of a sale and if the entire project is considered, top and bottom prices (Table 6.6) are used to calculate the internal rate of return (IRR) (21.8% and 15.6%). The payback period (PP) is 9 and 10 years and the net present worth (NPW) LIT 206.8 bn and 114.9 bn. The alternative hypothesis, management, gives 20.70% and 15.7%, a PP of 10 and 11 years, and a NPW of LIT 232.5 bn and 143.9 bn. Even in the more conservative estimates, with an IRR equal to 15.7% in real terms, it is higher than the average capital investment rate on the market. Results are positive in terms of net profits, ranging between a maximum of LIT 232 bn and a minimum of LIT 115 bn for a total investment of LIT 280 bn and a return over 9–19 years.

According to the economic feasibility study carried out by the Turin municipality, the increase in the market value following the conversion can be estimated at LIT 46.4 bn in the more favourable hypothesis. The analyses carried out for each function are even more interesting, since they highlight different profitability levels according to function.

The exhibition centre appears to offer the best profitability level: it is a strong function, the pivot of the entire operation. The total cost of LIT 62.4 bn has a standard market value of LIT 74.9 bn (equal to 120% of the building costs) and a marketing mix of 100% in the case of renting. Top and bottom IRRs are 35.9% and 28.6%, with PPs of five and six years, and NPWs of LIT 198 and LIT 138 bn. The conference centre has a cost of LIT 14 bn, with a standard market value of LIT 17 bn, corresponding to 120% of building costs. It has rather low financial feasibility indices, with 2.8 and 2.2 IRR, a PP over 20 years, negative NPWs with LIT 6.5 bn and 7.3 bn (top and bottom values).

The incubator would cost LIT 13.4 bn with a market value of LIT 24.8 bn (a sales price of LIT 1.5 bn/m^2) and a marketing mix around 100%. The PP would be 10–11 years and the NPW 5.9 and 7.6. Overall it gives an intermediate picture, placing itself in between the other two. The 'innovation and discovery centre' (*centro per l'innovazione*) has a cost of LIT 40.4 bn and a market value of LIT 59.9 bn (sale price LIT 1.8 bn/m^2). Profitability levels appear to differ according to the marketing mix. In the more favourable hypothesis (75% sale and 25% rent), the IRR would give 14.4% and 15.7%, a PP of five and six years, and a NPW of LIT 10.9 bn and 11.9 bn.

The university has a cost of LIT 61.3 bn, and a market value, taken to be the same as that for offices, of LIT 80.7 bn (sales price LIT 1.5/m^2), with an IRR of 21.6%, a four-year PP and a NPW of LIT 9.6 bn for the top price and a 14% IRR, with a 10-year PP and a NPW of LIT 37.4 bn in the event of the top price with 100% mix. Services and retail business have a forecasted cost of LIT 9.4 bn, and a market of LIT 17.0 bn (LIT 2.1 million/m^2); its IRR reaches 48%, and three-year PP; the new NPW is LIT 5.6 bn, with 75% sales and tenancies.

The residential centre costs LIT 17.7 bn, a market value of LIT 24.6 bn

(LIT 1.8 million/m^2). In the more favourable hypothesis, its IRR would be 18%, a five year PP and lastly a LIT 4.9 bn NPW, with 75% sales and 25% tenancies. Car parks cost LIT 53.8 bn, a market value of LIT 50.8 bn (sales price LIT 1.4 m/m^2), IRR between 6 and 8%, a negative NPW of LIT 2.3 bn in the best of all cases. The office block (the *palazzina*), is expected to cost LIT 4.4 bn and has been classed as a special project.

In evaluating the results of the economic feasibility study, one ought to bear in mind that Fiat does not monopolize the real estate market, as the size of the Lingotto might lead one to believe. Quite the opposite: if anything, the negotiation began under very unfavourable general conditions, especially in the initial phases. The premise had no other buyers both because of its size and because of the reorganization the whole industry was undergoing; furthermore, there was no demand for large industrial buildings. However, potentially there was a very strong demand for the approximately 6 000 000 m^2 of available industrial areas, if its use classification were changed. This was no easy feat if one considers the political and administrative climate of the time, and the impact the closure of the plant had had on the job market. It was a gamble, and this explains why Fiat has been so active in promoting the building, the aim being that of converting the building and using at least part of it for activities that might be classed as 'in the public interest'. One ought not to forget that in any case if the whole of the Lingotto were to be transformed, the size of the premises far exceeds existing demand in the private sector.

REFERENCES AND FURTHER READING

CENSIS Servizi (1988) *Il mercato e dilizio nelle aree metropolitane*, Rome.

D'Agostini, S. and Lisciandra, G. (1984) *Localizzazioni industriali*, Franco Angeli, Milano.

ISTAT (1988) *Le Regioni in cifre*, Rome.

Lassini, A. (ed) (1984) *La politiche regionali per la localizzazione industriale*, Regione Lombardia, Franco Angeli, Milano.

Sorrente, G. (1977) *Regioni industria territorio*, Franco Angeli, Milano.

Tosi, A. (ed) (1983) *La politica industriale della Regioni*, CLUP, Milano.

7

THE NETHERLANDS

Barrie Needham and Bert Kruijt

The aim of this chapter is to describe and explain the market for industrial property in the Netherlands. This is done in the following way. The workings of the industrial property market result in the provision of industrial land and industrial buildings in a certain quantity, in certain locations, at certain prices, etc. These can be regarded as the outcomes of the industrial property market, and it is these outcomes which are described in this chapter. But this chapter aims also to explain the outcomes. The explanation is sought by: identifying the actors involved; describing the context within which they take their decisions; describing the decisions which the actors take, and explaining these decisions as being the result of the interaction between the various actors mutually, and between the actors and the context.

The chapter is structured as follows. The outcomes of the industrial property market are described. The main actor in this market is described: this is the industrialist, who uses the property. He is the final user and it is his demand which leads to industrial property being constructed. The (other) actors involved in the industrial property market are identified (of which the actors in the public administration are mentioned only briefly: they can be identified in detail only after the description of public law and public policy).

The context is described within which demanders and suppliers interact with each other and take their decisions. The following aspects of this context are treated: the law relating to civil contracts, financial legislation, and the law with which the public administration can intervene in the market for industrial property. Public policy which has a direct effect on the industrial property market is described. This public policy aims to regulate this market and also to participate actively in it. Using the information in the above sections, an attempt is made to explain the outcomes as described in section 7.1. Finally, some case studies of industrial property development are described.

7.1 THE OUTCOME OF THE INDUSTRIAL PROPERTY MARKET

7.1.1 Industrial buildings, in total

No official statistics are published for industrial buildings as such, but we can derive estimates about the annual production and the stock of industrial buildings, as follows. The final user of industrial property regards it as a factor of production, a necessary input in his production process. The construction of industrial property is therefore an example of new capital formation in the productive sector. Statistics for new capital formation per year and for the value of the capital stock of industrial property can, therefore, be used to estimate the volume of industrial property. These statistics are not available for all users of industrial property; but they provide a starting point. (We must be aware that the acquisition of land is not regarded in economic theory as new capital formation; for some analyses, therefore, it is better to exclude expenditure on land.)

In the eight-year period 1980–1987, the annual average value of new capital formation by manufacturing, public utilities, mineral extraction, and the building industry all together was NFL 2304m in buildings and NFL 185m in land (Investeringen in vaste activa in de nijverheid 1988, CBS, Voorburg). We know, however, that many other types of productive activity use industrial property. Figures are available for industrial property bought or rented on the open market by all users. In the period 1983–1987, the average per year was around 730 000 m^2 for all users: of this, the amount bought or rented by manufacturing, public utilities, mineral extraction, and the building industry together was around 230 000 m^2 (=31.5% of the whole) (Hogenbirk and Loggers, 1988). The conclusion can be drawn that, if that proportion for acquisitions on the open market applies to all industrial property (and in particular to industrial property commissioned or bought by the final user), then the NFL 2304m a year which the four sectors mentioned above invested in new buildings in the 1980s can be multiplied by 100/31.5 to give the investment in (new) industrial buildings by all sectors (NFL 7300m).

This estimate is derived from statistics for the value of industrial property acquired by the final user (official statistics collected from 1988 onwards include industrial property rented and in leasehold ownership). But the final user might also lease industrial property from a property management company or other landlord. We want to know, therefore, how much industrial property is built for leasing.

Of all the industrial property bought or rented on the open market in the period 1983–1987, only around 70 000 m^2 per year (10% of the total) was new (Hogenbirk and Loggers, 1988). With an average rent of NFL 80/m^2 per year and an initial yield of around 10% (Zadelhoff, 1990), this represents about NFL 56m a year commissioned by others than the final users. (This value is of the same order as the estimate of NFL 36m in 1989 made by van Waveren, 1990.) The new property leased by final users is part of this NFL 56m, but so is the new property bought by final users from property developers: the value of the new property leased by final users must, therefore, be less than NFL 56m a year, which is itself a very small proportion (0.76%) of the value of new industrial property acquired (bought or commissioned) by the final user. This can be compared with the office market, where – it has been estimated – about 40% of the annual production in square metres is built for the market and not commissioned by the final user (Lukkes *et al.*, 1987, p. 10).

Our estimate of the value of industrial property built per year (NFL 7300m) must be regarded as very rough; but it serves as an order of magnitude for the rest of this chapter. And its significance for the national economy can be seen not just from its absolute value but also from its part (=9.6%) of all new capital formation in the economy (again excluding land).

Estimates of the size and value of the stock of industrial property are more difficult to find or make. Blankenstein-Bouwmeester and Lukkes (1984, p.1) give a figure of 40m m^2 but it is not supported by any argument; another publication (VROM 1990, p. 22) gives precisely five times that amount! For manufacturing industry (but no other productive sectors) official statistics give the value of the total stock of land and buildings at the beginning of 1987 as NFL 75.3bn (buildings) and NFL 17.5bn (land).

7.1.2 Industrial buildings, by user

Official statistics for the value of industrial property acquired per year by user are available only for the manufacturing sector, public utilities, mineral extraction, and the building industry: for 1987 these values were (buildings only) NFL 2513m, NFL 295m, NFL 56m, and NFL 234m respectively. The value of industrial property acquired by other sectors is not known. The exception is that (very small) proportion of industrial property bought or rented on the open market. There, the value per sector was (annual averages 1983–1987, source Hogenbirk and Loggers, 1988): manufacturing, public utilities, mineral extraction together, about 200 000 m^2; the building industry about 30 000 m^2; warehousing etc. about 260 000 m^2; transport and storage about 110 000 m^2; the construction industry, non-financial commercial services, and public administration about 60 000 m^2 each.

7.1.3 Industrial buildings, by location

Anything we can say about industrial property by location is again restricted to what we can derive from the two sources used above (Investeringen in vaste activa in de nijverheid 1988, CBS, Voorburg; Hogenbirk and Loggers, 1988). For all industrial property acquired, but only by manufacturing industry, concentrations were to be found in Twente (3.5%), Arnhem/Nijmegen (13.5%), Utrecht (5.7%), Greater Amsterdam (4.3%), Greater Rijnmond (4.6%), north-east of North Brabant (4.1%), west and south-east of North Brabant (16.7%), southern Limburg (8.2%) (buildings only, 1987).

An analysis of industrial property acquired on the open market between 1983 and 1987 (Hogenbirk and Loggers, 1988) gives a somewhat similar picture of geographical distribution and includes all sectors. This shows a concentration around Amsterdam, Rotterdam, Utrecht, Eindhoven, but also the IJsselmeerpolders and the Haarlemmermeer, south-west Brabant, 's-Hertogenbosch and Twente. That analysis is for all users who bought or rented industrial space. The same study made a separate analysis of those who bought or rented in the trading and transport and storage sectors. Users in the first sector were to be found in large numbers in Amsterdam, Rotterdam, Eindhoven and The Hague, around Schiphol airport, in the IJsselmeerpolders and to the west of Utrecht; users in the second sector were more concentrated, in particular around Rotterdam and Schiphol, but also between Utrecht and Amsterdam.

7.1.4 Industrial buildings: who commissions and owns them?

The calculation made above gives a clear answer to this question. Almost all industrial building is commissioned by the final user: the proportion commissioned by a property developer for sale or rent is tiny.

An indirect confirmation of this is given by the following calculation. In the four years 1987–1990, around 740 000 m^2 of industrial property were rented or sold on the open market per year (Zadelhoff, 1990). The total area of industrial property in 1984 was variously estimated to be either 200m m^2 or 40m m^2 (see section 1.a): the annual turnover represents, therefore, either 0.37% or 1.85% of the stock. Whichever is the correct figure, it is small: for example, a household spends, on average, around 10 years in one dwelling, giving an annual turnover equal to 10% of the stock. The explanation must be that most industrial property is never transacted on the open market, but is commissioned by the final user who uses it until it is written off. We can conclude that almost all industrial buildings are owned by their users; and even more than are commissioned by their users, for some of the buildings commissioned by property developers will subsequently be bought by the final user. A very small stock remains which is rented to the final users. This stock will be held, mainly, by the 'financial institutions' and other property-management companies.

The Dutch financial institutions own very little industrial property. They have proportionately less real estate in their portfolios than their British counterparts, preferring fixed-interest securities (de Geus, 1990). And when they do invest in real estate, they prefer offices. An analysis of the portfolios of some of the largest funds (Algemeen Burgelijk Pensioenfonds, Nationale Nederlanden, Shell, Bogamij) found that a very small proportion of the value was in industrial property (Rouwenhorst, 1991).

This is a good place to point out that a new type of industrial building has been introduced on to the market, which includes both industrial and office space. This is a recent development, and it is unclear to what extent it is included in the statistics for industrial property used above. However, it is proving very popular with property developers: their construction costs for such buildings were NFL 77m in 1986, and this had risen to NFL 222m by 1989 (van Waveren, 1990). One such building has been chosen for a case study in section 7.7.

7.1.5 Industrial buildings: prices and price changes

It follows from the above that there is only a small active market in industrial property (i.e. a market in which prices can be set). Figures are, however, known for rents realized in transactions on the open market for industrial property. These show that rents were at 'dump' levels for several years, but are now beginning to grow, especially in Amsterdam and Rotterdam. Levels in 1989/90 were on average around NFL 80/m^2 p.a., with little variation around that average: as low as NFL 40 in the peripheral regions of the north and east, rising to around NFL 100 in parts of the Randstad and NFL 160 at Schiphol (Zadelhoff, 1990).

7.1.6 Industrial land

Every year, the National Physical Planning Agency (*Rijksplanologische Dienst*), together with the Ministry of Economic Affairs, carries out a survey of industrial land. All industrial estates on which at least 5 ha are still available are covered, and the results are published for each estate separately describing:

1. how much was disposed of the previous year;
2. the area still available;
3. of that latter, the area immediately available (i.e. already serviced);
4. the price (sale or ground lease) per m^2.

The following results are relevant for this chapter (see report in *VastGoedMarkt*, 1989). The total area disposed of per year declined steadily until 1982, since when it has been rising steadily (in 1988, it was 1173 ha). Nevertheless, the available supply is still generous and, in some parts of the country, wildly excessive. If disposals continue at the same rate and no new land is supplied, the province of Utrecht is in the best position (supply sufficient for another five years), the province of Groningen in the worst (supply sufficient for another 80 years!). Where the supply is most excessive, it is mainly of industrial land with direct access to ports. The provinces where, in the last three years, most industrial land has been disposed of are North Brabant (far and away the most), followed by Gelderland, South Holland, North Holland. Prices vary by location, as would be expected. The variation is not, however, great, with most prices lying between NFL 40 and NFL 100/m^2. It is only in exceptional cases (e.g. around Utrecht, Amsterdam, Schiphol, The Hague, Rotterdam) that prices rise above NFL 100/m^2.

Those facts prompt the question: who is responsible for supplying that industrial land in such a prodigal way? The answer is mainly the municipalities, with the harbour sites provided by the (public) port authorities. In the six-year period 1978–1983, of all industrial land supplied for building upon, 69% was provided by the municipalities, and of the rest, a lot would have been provided by the port authorities (Maandstatistiek Bouwnijverheid: unfortunately, these figures have not been collected after 1983). Why municipalities do this is described in section 7.5.

7.2 THE FINAL USER AS DEMANDER OF INDUSTRIAL SPACE

Industrial property is built, in the final analysis, because it is to be used for productive purposes. The final user – and therefore the demander – is the industrialist. The market for industrial property cannot be understood without an understanding of changes in industry: how the composition of the industrial sector is changing, how the location of industrial production is changing, and how the requirements which industry has for production space are changing. In this section we sketch these changes. Once again, however, our task is made difficult by the fact that most statistics use a definition of industry that is different from that used in the industrial property market.

7.2.1 Industrial property as a producer good

What does the final user want from industrial property? First, he wants space for his production processes, in sufficient quantity and with the right properties. These properties might be highly specific to his production processes: he wants a building closely tailored to his needs. Second, he wants to be able to finance this in a way which is most convenient for himself. If he buys the property, this requires capital. His capital and his opportunities for acquiring more are not unlimited: so buying industrial property is usually at the expense of some other use of capital, another use which might be more productive (e.g. research and development, advertising, machine tools). So in many cases, the industrialist will want to rent his production space.

We can see already that these two demands – for industrial space designed for one's own production processes, and for renting that space – can be incompatible. For a landlord will acquire property in order to rent it out only if there is a pool of potential users; and if a user demands space with properties which other users do not want, then no landlord will take the risk of acquiring for rent. We come back to this point later.

Other aspects which the final user wants from industrial property we can deduce from the following figures. Investment by manufacturing industry in industrial buildings (NFL 2512m in 1987) represents 3.3% of the value of its stock of industrial buildings in 1987. This can be compared with the construction of new dwellings in that period, which added an average of 2% to the stock in 1987 – but measured by volume, not by value. One can conclude that industrial buildings have a shorter economic life than dwellings. In 1987, new industrial buildings represented 16.2% (and land another 1.4%) of all new capital formation by manufacturing industry. In that same year, the stock of industrial buildings and land represented 38% of the value of all capital stock owned by manufacturing industry. Two conclusions follow from this.

1. Industrial buildings have a longer economic life than other capital goods owned by manufacturing industry.
2. The decision to invest in industrial property is just one of many investment decisions made by manufacturing industry and certainly not the most important.

Financially, the acquisition of land represents only 1.4% of all capital investment by manufacturing industry, which puts into perspective the political importance often given to 'competitive' prices for industrial land. It also puts into perspective the argument that an industrialist will be happy to take land on a ground lease rather than buying it, because buying it ties up his capital.

7.2.2 The composition of the industrial sector

One indication of demand for industrial property is given by the absolute increase in the number of workplaces (although this relates only to the demand for extra space, not for better – replacement – space). Of those sectors which acquire general-purpose industrial property on the open market (see section 7.1), between 1982 and 1988 the number of employed grew substantially only in warehousing (+26 000) and transport and storage (+49 000); the building industry grew less (+16 000) and in the second half of that period (1984–1988) engineering grew by 23 000 after a decline. It is in these sectors, therefore, that demand for industrial property would be expected.

Another indication is given by the absolute increase in turnover. Within the manufacturing sector, this was to be found (1980–1987) in paper and paper products (+NFL 2.2bn), printing and publishing (+NFL 3.8bn), mechanical engineering (+NFL 2.0bn), machine goods (+NFL 3.6bn), and electrical (including electronic) engineering (+NFL 6.1bn). An analysis of absolute increases in value added over the same period shows the biggest growth in the same manufacturing sectors, with electrical and electronic engineering showing a much greater absolute increase than the other sectors. This conclusion is supported by the figures for the types of user acquiring industrial property on the open market (see section 7.1).

7.2.3 The location of industrial activities

These analyses of changes in employment, turnover and value-added indicate which industrial sectors will be demanding industrial property. But where is that industrial property being demanded?

Statistics are published for the number of workplaces per province in the various industrial sectors, and the absolute increase in workplaces can be used as an indication of the demand for extra (not replacement) industrial property. The general conclusion drawn from these figures by de Smidt and Wever (1987, p. 70) is that the 'centre of gravity' of manufacturing industry is moving south: growth is strong in North Brabant, Zeeland and Limburg. If we extend the analysis to include two of the sectors outside manufacturing but which nevertheless use industrial property, then we see the following. In the trade sector (which includes warehousing) the absolute increases between 1980 and 1986 were high in North Brabant (+18 000), Gelderland (+15 700), South Holland (+10 700), North Holland (+9600), Utrecht (+7700). In the transport and storage sector, growth was high in North Holland (+14 200), South Holland (+11 800), North Brabant (+5600), Gelderland (+4400) (all figures from the National Office of Statistics).

Firms demand industrial space when they are new, so an analysis of where firms are started will give yet another indication of the location of demand for industrial space. It appears (de Smidt and Wever, 1987, pp. 114–117) that the cities of the Randstad generate many firms; also North Brabant and Middle Gelderland. The Randstad is more important as a starting place for firms in the service sector than in the manufacturing sector. The 'half-way zone' (between the Randstad and the peripheral areas) is the birthplace of many firms in the manufacturing sector, as is Rijnmond (around Rotterdam and Europort). It is the southern part of the half-way zone (North Brabant) which accommodates the new firms orientated to international markets.

Analysis at a smaller geographical scale complements this picture. The growth in manufacturing employment in the Randstad mentioned above did not take place in the core cites (Amsterdam, Rotterdam, The Hague, Utrecht) but on the fringes of those cities: an extreme example of this is the strong growth around Schiphol airport (van der Vaart and van Weesep, 1987). The absolute decline of manufacturing employment in the core cities does not bode well for the success of current policy to renovate industrial estates in those cities (Ruimtelijke Verkenningen, 1987, p. 31): how strong will be the demand for industrial space when the redevelopment – heavily subsidized out of public funds – has been completed?

7.2.4 The type of industrial space demanded

The activities which require industrial space are undergoing significant changes affecting both the choice of what to produce and of how to produce it. There is diversification and innovation in what is produced; and the application of computer technology is affecting the production process. An important consequence is radical changes in the organizational structure of the firm. This relates not only to direct production, but also to the totality of activities within the firm: a reduction of the number of ranks in the hierarchy, much shorter channels of communication, and much faster reaction to external changes.

These changes are affecting the requirements which the users place upon industrial property. The following effects have been noted in research into engineering firms (Laurier *et al.* 1987): it is probable that there are already being experienced, or shortly will be, in other sectors also. Changes in what is produced are causing a breakdown in the traditional division of space into that for direct production (making and assembling) and that for indirect production (organizing, management, design, etc.). Changes in production processes are causing industrial space to be demanded of a different quality and quantity. Changes in quality refer to the production environment (temperature, humidity, dust-free, vibration-free mountings etc.), and to the appearance of property as an advertisement for the firm itself. Changes in quantity refer to the fact that less space per unit production is being demanded. One reason is that production is better organized and therefore quicker; another is that better logistics leads to fewer stocks being held; another is that the space is used for more hours a day.

The same research found changes also in the requirements placed by manufacturers on the location of their production. These included high-quality surroundings, as an advertisement for the firm, and proximity to suppliers, customers, and subcontractors. The latter might be an unexpected finding: the whole development of industrial society has been an increasing division of labour and regional specialization. However, the technical and organizational changes noted above are leading to much more frequent and intensive links between the different establishments involved in the production process. Holding fewer stocks means more frequent deliveries; keeping technically up to date means frequent exchange of ideas; supplying to a rapidly changing market means close contacts with customers; demanding high quality means strict quality controls on suppliers. The research suggests that, instead of a region specializing in one industry or in one stage of the 'product life cycle', each region will contain networks between all the establishments in one production chain and between all the stages of the product life cycle. Certainly, analysis of the changes between 1963 and 1985 shows that regional specialization had already begun to diminish (de Smidt and Wever, 1987, p. 62).

7.3 THE OTHER ACTORS: SUPPLY

There are many actors involved in the market for industrial property besides the final users, and the more important ones will be mentioned here. However, for most of them, nothing more than a mention is needed, for their part in the whole is neutral.

One of the most important is the public administration, which is only dealt with briefly in this section because it is treated separately in sections 7. 4 and 7. 5. The public administration has a triple role in the industrial property market. It tries to regulate it, for which it uses certain legislation. This legislation is the subject of section 7. 4, where the public agencies involved in applying it are mentioned. The ways in which the legislation is applied (the policies pursued with that public legislation) are the subject of section 7. 5. The second role of the public administration is a more active one: it supplies industrial land (and occasionally industrial buildings too). This also is described in the section on public policy. The third role of the public administration is as final user, demander of industrial property. We have already seen in section 7. 1 that the public administration acquired on the open market on average 60 000 m^2 per year, and public utilities another 60 000 m^2 per year (1983–1987). We need, however, pay no further attention to this third role, for there is no reason to think that the demands which the public administration places on industrial property are different from the demands of other final users.

The people involved in the construction of industrial property are the building contractors, with all their advisers (quantity surveyors, architects, structural engineers, etc). Their influence is neutral. They are commissioned to build by, in most cases, the final users (see section 7. 1). Others who commission industrial property, but not for final use, are property developers and, in a few cases, public agencies as part of their policy for the local economy (see section 5). The property developer *per se* (i.e. one who does not retain the building once completed) builds only those properties for which there is a large and ready market. He sells either to a final user or to a landlord (property management company, etc.). A landlord too is interested only in that property which can be rented to a large and ready market.

Another source of supply is of second-hand industrial property: property which was not previously available on the open market (it was occupied by the owners). If that owner no longer wants to use the property, he will consider demolishing it and rebuilding on the site, or renovating it, or selling it. An important consideration in that choice is the price he can expect to realize if he sells. If the building was designed to meet his own specifications, then it will have a low resale value and the owner might prefer to rebuild or renovate. Also possible is that the building itself has a low second-hand value but the site on which it stands has a high (re-development) value: the property is then supplied second-hand but not for re-use.

The construction or acquisition of industrial property must be financed: the financiers play a neutral role, but they can be influential in two situations. If the final user wants to buy his property but cannot borrow the money, he will consider renting (see section 7. 2). And if an institutional investor acquires the property, then it has special requirements as follows. Institutional investors in the Netherlands are anything but short of money to invest: but they have many openings for that money. They can buy shares, they can buy fixed interest bonds, or they can buy buildings (including industrial buildings). When making that choice they look for security, good returns, liquidity, and a spread of risks. They might want to invest in industrial property in order to diversify their portfolio; but apart from this passive reason, would they acquire industrial property for its security, returns, or liquidity? Users of industrial property are vulnerable to fluctuations in the economy: in any case, the volume of new industrial building fluctuates more than that of offices and shops (van Waveren, 1990; Blankenstein-Bouwmeester, 1984, p. 56) This reduces the security of income from rents. However, the industrialist will not move readily from a property which meets his specific requirements: and that makes him a more secure tenant.

Institutional investors must receive returns to meet their obligations (e.g. paying out pensions), and they charge premiums on the basis that a return of 4% (de Geus, 1990, p. 16) will be just sufficient. The returns to the landlord from industrial property are rent and capital growth. If the rent is stable, then the returns are unattractive. Capital growth comes if rents also grow and/or if profitable re-development of the site is possible. In the Netherlands, rents for industrial property have been low and stable. Re-development becomes profitable if the site becomes more desirable as a location during the economic life of the building, but that requires great variation in the locational characteristics of industrial sites and in that respect the Netherlands is a fairly uniform country. The outcome is that the financial institutions are not very interested in industrial property. Quite a lot was built for rent in the 1970s, but when demand dropped as a result of the economic depression, rents stopped rising and the financial institutions were not interested in acquiring more industrial property. It is only recently, with the introduction of mixed office/industrial buildings, that industrial property has become an attractive investment again in certain locations, with rents rising and initial yields falling (Jones Lang Wootton, 1989/90). Rented property is a liquid investment if it can be sold easily with a sitting tenant, or if it can be sold relatively easily after falling vacant. Neither of those conditions will be met if the building has been designed (or adapted) to satisfy the specific needs of the final user.

The final group of actors to be mentioned are the professional advisers, in particular the estate agents who specialize in selling or letting industrial property, and the valuers who have to assess its value. In the Netherlands, the professional title of broker in real estate is protected by law. Most brokers are members of the *Nederlandse Vereniging van Makelaars*, which has a little over 2000 members (1991) and also a section for those who want to specialize in commercial real estate. Training for that profession is organized by the professional body itself, at an academic level below that of the universities. There is only one university course in real estate, a post-graduate part-time course at the University of Amsterdam.

Industrial property has to be valued for local property tax (see section 7. 4) and some other statutory purposes. Because so little industrial property is transacted on the open market, there are very few market guides to the value, so assessments are usually based on replacement value.

7. 4 THE LEGAL CONTEXT

7. 4. 1 The law relating to civil contracts

By this is meant that legislation which applies to all legal persons, including public agencies, when they enter into voluntary contracts.

Rights and interests in land and property

Civil contracts relating to land and buildings are regulated in principle by the general law regarding rights of ownership and use, of both movable and immovable property. This law is anchored in the Civil Code (*Nieuw Burgelijk Wetboek*) and gives the owner of landed property the same rights (in principle) as the owner of a motor car, a painting, or a fountain pen! In the case of land and buildings, the legal owner may grant to others rights to use his property. If these rights are particularly extensive and durable, the enjoyer of them is regarded legally as having economic ownership of the property (as distinct from legal ownership). This is a recent development: it is not based in civil law but has grown out of taxation law (see section 7. 4. 2) and was legally recognized for the first time in 1955. The person who has the economic ownership has the complete economic interest in the property (including the risk associated with changes in value). Also, the owner of the economic rights has to include the property as an asset on the balance sheet, and may write off the value of the ownership against tax. The legal title remains with the vendor, for a period to be agreed upon, after which it is transferred to the economic owner. If, in the meantime, the economic owner sells that right, the legal title may be transferred to the new economic owner, who however has a weaker right to it than the first economic owner. Rights which are less extensive and durable (e.g. rights of tenancy) are regarded as user rights, not ownership rights.

Further, all rights can have the form of a real right (*zakelijk recht*, rights *in rem*, which is attached to the property and can, therefore, be freely traded) or a contractual right (*persoonlijk recht*, rights *in persona*, between the legal persons named in the contract). The most important real rights in land and buildings are ownership, servitudes or easements, ground leases, building rights (i.e. to own a building on land owned by another), and condominium rights. An example of a personal right is a tenancy to rent. These distinctions are particularly important for financing and taxation (see section 4 below).

Finally, rights of ownership may be encumbered by servitudes, consisting of an obligation to allow something to take place on another property or to desist from doing something on one's own property. Such encumbrances are legally specified in particular cases and are additional to the general restrictions on the rights of ownership, consisting of the obligation not to cause a nuisance to adjacent properties.

Landlord/tenant legislation

The relationship between the landlord and the tenant of industrial property is determined primarily by the contract to which both voluntarily bind themselves, and the content of that contract is primarily a matter for the two parties alone. That is the principle laid down in the Civil Code (*Nieuw Burgelijk Wetboek*). The only other legislation regulating this relationship is in the Rent Act (*Huurwet*) and applies only to developed land. The level of rent is purely a matter for the two contractual partners. Common practice is to agree to an inflation-linked rent plus rent-revision after five years if the rent has grown out of line with market levels. The contract usually also incorporates provision for paying service costs.

The expiration of the lease is in itself not sufficient to oblige the tenant to quit the property; the landlord must request that separately and explicitly. The tenant may request the court to postpone the obligation to quit, for a maximum of one year, but the request can be repeated twice. This obligation can be postponed if the court considers that moving would damage the interests of the tenant more than his not moving out would damage the interests of the landlord. Appeal against the court's decision is possible only on a point of law. The lease itself cannot be extended by a court order: if the obligation to quit is postponed, the court can fix the compensation (in lieu of rent) which the tenant must pay to the landlord (Smit, 1989,) Note that this description applies to industrial and office premises, not to dwellings, nor to commercial property providing services directly to the public, such as shops and restaurants.

The disposal of land

The general civil law relating to the rights of ownership (see above) govern also the disposal of that right. There are two aspects of this which are important for this chapter, and it is important also to know that they apply equally to public and private bodies disposing of land.

The first aspect is that land may be disposed of under various legal constructions, of which the only ones common for industrial land are selling ownership of the land outright and selling a building (ground) lease on the land. The second important aspect is that, when disposing of land, it is not only restrictive covenants which may be imposed (specifying how the land may not be used) but also positive obligations (such as that the person buying the land must maintain the building up to a certain level, or must contribute annually to the costs of communal facilities). A supplier who chooses to dispose of land leasehold can use this so as to have a stronger and longer-lasting control on the use of that land. To a lesser extent, the power to impose positive obligations when land is sold can be used for the same purpose: it is legally possible, when selling the outright ownership, to impose conditions which bind not only the first buyer but all subsequent buyers (*kettingbeding*).

7. 4. 2 Financial legislation

Financing

This will be considered from the point of view of the user of the industrial property. He can own it outright, including the land, financing the purchase either from capital or with a loan (usually in the form of a mortgage). If the building is on land disposed of by ground lease, then he must finance the construction costs: the costs of the ground lease he will pay either periodically or as a lump sum (a premium). Payment of ground leases can be set against income tax obligations. Finance for a premium can be borrowed by means of a mortgage. At the other extreme, the user can rent the property, paying a periodic rent. The rights of ownership (both legal and economic) rest with the lessor; the lessee has certain specified rights of use.

There are a number of other possibilities, which can be regarded as intermediate between owner-occupation and renting. One of these is *huurkoop*, literally 'hire purchase'. The user pays a sum periodically for a number of years to the legal owner under a contract whereby the user becomes the legal owner when the pre-arranged payments have been completed. It is a sort of deferred purchase. As a method by which the user finances his occupation of real estate, this is not very common. Another construction is the finance lease, whereby the lessee acquires economic ownership at the beginning of the lease; the lessor retains legal ownership during the lease but subject to the lessee using his option to acquire the legal ownership every (say) five years. At the expiration of the lease, the lessee has the right to acquire the legal ownership for a nominal sum. The lessee cannot terminate this contract during its course. A third construction is the operational lease. With this, the lessor has not only the legal ownership but retains the economic ownership also, even after expiration of the lease contract. However, the lessee has, during the period of the lease, the option to buy.

To the user, the advantage of leasing in comparison with owner-occupation is that he does not need to finance the property out of capital or with a loan (which, if it is a mortgage, will usually not cover 100% of the costs, leaving the user to find the rest out of capital): the costs of the property can be paid out of operating income. The advantages in comparison with renting are that the outgoings are fixed at the beginning of the lease period and cannot fluctuate, as rents can (usually rising!); also, the user can determine at the beginning of the lease what he will do (buy or vacate) at the end of the lease.

The most advantageous to the user would seem to be the financial lease: so why is not more use made of it? The reason lies in the unintended effect of the *Tijdelijke wet huur-koop onroerend goed*. This law was passed in 1973 to protect people who acquired housing in this way and it provides that the lessee can redeem the loan at any time. This possibility is not attractive to the lessor. If he has financed the lease when interest rates were high and rates then fall, the lessee might want to pay off the loan; the lessor then has a fixed interest obligation that costs more than he can get by re-lending the money. It has been interpreted that this Act applies also to business premises held on lease or *huurkoop*: so the user of industrial property wanting to enter into such an agreement often cannot find a willing lessor.

Taxation

In the law regulating value added tax (BTW), real estate has a special treatment, because it is such a durable good. If property is sold within two years of its completion, the seller has to pay 18.5% BTW. This applies also to land which has been newly serviced (see section 7.5). The buyer will usually be able to claim back that BTW. If he should then resell the property, no BTW would have been paid on it: so the rule is that if it is resold within nine years after completion, the seller must repay a part of the BTW repayment he received at the time of purchase. This BTW ruling applies to the acquisition of both legal and economic ownership and thus, for example, to the acquisition of financial leases, but not operational leases. When property is rented, the lessor does not have to pay BTW on the rent received. But then he cannot set off operating expenses against his BTW obligations.

When real estate is transferred and BTW does not have to be paid, then a transfer, or conveyancing, tax (*overdrachtsbelasting*) has to be paid, at 6% of the economic value: but not if the property is resold within three months, a provision which some property speculators take advantage of! This is paid by the buyer, and it applies to the transfer of rights of legal ownership (not economic ownership). This explains the increasing use of economic ownership: it avoids the liability to transfer tax. If the person who acquires economic ownership later exercises his right to acquire legal ownership, at that later date he will be liable to the transfer tax. Transfer tax is liable also when commercial user rights (easements, ground leases, *usufruct*, etc.) are exchanged.

If property is held as a business asset, then any profits or capital gains made when it is sold are liable to company income tax (*vennootschapsbelasting*), unless the proceeds are deposited in a replacement reserve, where they can stay for up to four years. Some types of company are not liable to company income tax. These are the institutional investors: for them investing in property is more advantageous than for other companies. No tax is due on profits (capital gains) made when the property sold was being held as a personal possession. Income received from rent is always liable to tax, whether from property held for business purposes (company income tax) or as a personal possession (personal income tax).

The user of a rented building can offset against tax the rent paid: the user of his own building, but also the lessee of a financial lease, can deduct against tax the interest payments on loans for the property and also the depreciation of the building. Ground rents paid under a ground lease are also tax-deductible.

There is a net wealth tax on individuals of 0.8% pa.

7.4.3 Relevant public law

By this is meant that legislation which enables public agencies to impose obligations on private persons and agencies. Before this legislation is described, it is useful to give a sketch of the structure of the public administration in the Netherlands.

The Netherlands is a decentralized unitary state with a system of constitutional parliamentary monarchy. Decentralized means that there are government bodies not only for the whole nation but also for every province and for every municipality. There are a few other bodies with governmental powers, such as the Water Boards. Unitary means that all those government bodies must work together. Constitutionally, this is realized by the national government's allowing the provincial and municipal governments to act only in ways specified by legislation passed nationally, and by making the provincial government subordinate to the national government, and the municipal subordinate to both provincial and national governments. In practice, municipalities in particular enjoy a lot of autonomy and have large budgets at their disposal.

The country is divided into 12 provinces, each with its own governing bodies, the Commissioner of the Queen, the Deputees-General (here called the Provincial Executive), and the States-Provincial (the Provincial Council). The governing bodies of the 700 Dutch municipalities consist of a mayor (*burgemeester*), a municipal executive (council of mayor and aldermen), and a council (*gemeenteraad*: see further Alders, 1991).

Physical planning

In this section, a very brief account will be given of the system for making and implementing land-use policy, in so far as it is important for industrial property. A brief account is all that is necessary, because in practice the control over industrial development exercised through the statutory system of town and country planning is light: the reason is that the municipality has other control mechanisms, which will be described in section 7.5.

A good starting point for describing the way in which physical planning sets a framework for industrial development is from the person who wants to build some industrial property. How is his freedom to do this influenced by the planning system? The crucial point is that no building works may take place unless first authorized by a building permit (*bouwvergunning*) regulated by the Housing Act, not the Physical Planning Act. So the would-be developer has to apply to the Municipal Executive for a building permit. This application is tested against three sorts of standards:

1. Does it comply with the building regulations (the technical regulations relating to structural safety, etc.)?
2. Does it require a permit under the Protection of Cultural Monuments Act or under a provincial or municipal ordnance?
3. Does it conflict with the local land-use plan (*bestemmingsplan*, b.p.)?

If the application fails on any one or more of these grounds, then it must be refused. The Municipal Executive must give its decision about the application within two months, which period can be extended once by at most two months.

The building regulations (the first test) must contain a clause that the appearance or visual aspects of the proposed development be tested: that is done by an independent body of experts (*welstandscommissie*) which advises the Municipal Executive. The third test allows the application to be tested against the land-use policy: it is, therefore, akin to development control, but there is no separate planning application as such. There are two exceedingly important aspects of this test against land-use policy:

1. if there is no valid b.p. in force (or if the municipal council has not taken formally the decision to start preparing one), the 'planning' test cannot take place;
2. if the application conforms to the b.p., then the application must be granted, and if the application does not conform to the b.p., then it must be refused (this is expressed by saying that the b.p. is legally binding).

The would-be developer of industrial property must, therefore, know three things: is there a valid b.p. covering the land on which he wants to build? If so, what is its content and who determines this? And how flexible is this legally binding document? As regards the first question, a municipality is obliged to make a b.p. for that part of its territory falling outside the built-up area. Within the built-up area, a b.p. is not obligatory, but most municipalities have made such plans for part or all of this area. As regards the second question, the content of the b.p. can be constrained in the following ways.

1. It must be approved by the Provincial Executive.
2. The municipality may choose to make a *structuurplan*: a land-use plan for part or all of its territory. This plan is not legally binding and its main function is to set a policy framework for the b.p. It is not a legal requirement that the b.p. be consistent with the *structuurplan*: but if it is not, then the chances of the b.p.'s being approved by the province and endorsed by the Municipal Council are reduced.
3. The province may choose (and all do) to make a *streekplan*, a land-use plan for part or all of its territory. If such a plan is in force, the province, when deciding whether to approve the b.p., will take into account its conformity with the *streekplan*. The Provincial Executive has the power to impose a directive (*aanschrijving*) upon a municipality in order to make the b.p. conform to the provincial planning policy.
4. The national government may take 'physical planning key decisions' about some aspect of the physical development of the country. A b.p. is seen by a regional agency of national government before being approved by the province: if this agency considers that the b.p. is not in conformity with national planning policy, it will advise the

province accordingly. It can happen, nevertheless, that the municipality wants to do something inconsistent with national policy: for example, when that policy requires that the municipality amend its b.p., the municipality takes no steps to do that. In that sort of case, national government can, with a directive, impose a change on the b.p., although this is a power which it uses with great reluctance and very rarely.

The third question is about the flexibility of the b.p. The statement 'a b.p. is legally binding' can give the impression that it is very inflexible. However, flexibility can be built into each particular b.p. Also, the system itself has a certain flexibility, irrespective of the particular b.p. (Verhagen, 1989). Flexibility can be built into a particular b.p. in five ways, as follows.

1. The plan can be drawn up in global terms, with the filling-in of the details delegated to the Municipal Executive.
2. The plan can be drawn up in global terms, with no obligation on the Municipal Executive to work it out in detail.
3. Certain aspects of the plan can be amended, the aspects and the scope for amendments being specified in the plan.
4. The Municipal Executive can be authorized to grant small exemptions from certain specified aspects of the plan.
5. The Municipal Executive can be authorized to impose conditions additional to those specified in the plan: these additional conditions may only be slight. This power increases the flexibility with which the municipality can implement the plan, but not the flexibility with which the citizen may deviate from the plan!

The whole system of statutory town and country planning has, in addition, certain components which may be used to give flexibility. These are as follows:

1. The b.p. can be revised or withdrawn. The whole plan must be revised after 10 years: but partial revision is also possible and can be used as follows. An application is submitted, which is inconsistent with the b.p.; nevertheless, the municipality wants to be able to grant it; so the municipality decides on a (partial) revision. The procedure, however, lasts several years.
2. The Municipal Executive can grant exemption from the conditions in the b.p. for a limited number of years. This can be used to allow temporary uses.
3. The Municipal Executive can also grant exemption permanently, but

only if the municipality is in the process of making a new b.p. to replace the existing one and if the proposed development would be consistent with the plan being prepared. This clause in the legislation (article 29) lends itself to abuse, and has in fact been widely abused by municipalities which wanted to approve development which the valid b.p. would forbid. In the 1985 revision of the Act, the procedures around this article were tightened up, and the extra flexibility (to achieve which that article had been abused) was created in a safer way.

4. If the applicant has his application for a building permit refused, he may appeal to the Municipal Council against this decision. Thereafter, he has the option of appealing under the Administrative Justice Act. Appeals are rarely used as a way of getting permission for development which contravenes the b.p.

(For an extensive treatment in English of the statutory town and country planning system see Bruil *et al.*, 1987; Dekker, 1991; Needham, 1988; Spit, 1987.)

If the application is for industrial development within a designated urban renewal area, then the application will be tested against an urban renewal plan instead of the b.p., and the powers and procedures are slightly different. If the application requires withdrawing a building (or part of it) from residential use, and it is in a municipality which has regulated this in its building regulations, then an additional permit (*woningenonttrekkingsvergunning*) is required.

Environmental protection

Most proposals for industrial development are subject to a statutory system of controls to enforce environmental protection, as follows.

The Public Nuisance Act (*Hinderwet*) requires that activities which can cause danger, damage or nuisance must obtain a permit to operate, to expand, or to change the operating processes. Most industrial production falls under this. In particular, air pollution by smaller firms is regulated by this Act. The Municipal Executive decides on the application. In an amendment of 1 November 1988 to the Nuisance Act, certain large industrial establishments have to prepare studies of the external risks they might cause, and they are obliged to report serious accidents. This is in accordance with the 'Seveso' Directive 82/501 of the EC.

The Air Pollution Act (*Wet inzake Luchtverontreiniging*) stipulates that certain firms must apply for a permit concerning gaseous discharges before opening a plant, expanding it, or changing its operations. The province is the responsible body. The types of firm falling under this legislation are the known (potential) air polluters (incinerators; storage of oil, coal, ore; etc).

The Noise Nuisance Act (*Wet Geluidhinder*) regulates noise from large industrial plants: these must apply for a 'noise permit' from the province. Also, the municipality must draw 'noise zones' around industrial estates: if these zones are drawn tightly, then they can restrict the development and expansion of firms on the industrial estate.

The Protection of the Ground Act (*Wet Bodembescherming*) aims to protect soil and ground water against pollution, and is implemented by the province. Moreover, it can overrule the Public Nuisance Act. A firm establishing in an area declared to be a 'water extraction zone' might have to take extra precautions to avoid dangerous substances coming into ground water.

The Surface Waters Pollution Act (*Wet Verontreiniging Oppervlaktewateren*) regulates discharge of waste water. All emissions into surface waters require a 'discharge permit', which can stipulate restrictive conditions. Further, an 'environmental tax' is levied on water discharges: for firms, the amount of tax depends on the amount of organic waste and, if the discharge of heavy metals is permitted, a tax is levied on this too.

Certain large-scale industrial activities are subject to environmental impact assessment. The results of this assessment are used to inform decision-making (including the decision about granting a building permit).

It will be seen that industrial development can be subject to a lot of environment legislation, some of which regulates only larger installations or certain selected types of industrial process. Some legislation, however, refers to (almost) all industrial development: this is the case with the Public Nuisance Act and the Surface Waters Pollution Act. The number of permits and agencies involved (especially with the larger installations) gave rise to co-ordination problems: it was to alleviate these that the Environmental Protection (General Provisions) Act was brought into force in 1980 (*Wet Algemene Bepalingen Milieuhygiëne*).

Other relevant public laws

Three other types of public law which can affect directly the industrial property market will be mentioned here.

A local property tax is levied on all real estate except land use for agriculture and forestry. The basis is the current value of the property and the tax is levied on both the owner and the user: if one person is both owner and user, he pays both taxes. (To give an idea of the size of this tax, the rates in the city of Nijmegen in 1990 were: for every complete NFL 3000 of value, NFL 8.15 to be paid by the owner, NFL 6.55 to be paid by the user; together, therefore, a property tax of 0.5%).

Until February 1988, under the WIR (*Wet Investeringrekening*) certain types of industrial investment (including new buildings) received a grant. The aim was to stimulate industrial investment in general. The working of the Act was such that an industrialist who commissioned a building for his own use was eligible for the grant, but since 1982 a property developer who built industrial premises for rent was not. This did not encourage building for rent.

Various supplements were available under the WIR, including one for industrial investment in designated development areas. In the course of the 1980s, however, the regional economic policy was weakened, and that supplement was withdrawn. One of the very few parts of regional policy still in force is the IPR (*investeringspremieregeling*), an investment grant available (also for new buildings) in areas with a structurally weak economy. It is now given in a few areas in the east of the Netherlands, and its highest value is 35%. It is not given for investment in buildings built for rent.

How the various public agencies use these public laws, often in combination with the laws applying to all legal persons (law on civil contracts) is the subject of the next section.

7.5 PUBLIC POLICIES

7.5.1 The provision of industrial land

In section 7.4 we said that, if the municipality wants to control industrial development, it has a more direct way of doing so than through the statutory planning system. We were speaking of the powers under the law of civil contract which the municipality has by virtue of being the principal supplier of industrial land. That most industrial land is supplied by municipalities is clear from the information provided in section 1. How municipalities do this does not need to be described here (instead, see Needham 1985a and Needham 1988). Why they do it is not simple, but one of the main reasons is that it gives the municipality a very potent set of instruments for guiding urban development in general and industrial development in particular. The powers available under the planning legislation are supplemented by the powers available to any supplier of land.

Those powers consist of:

1. being able to respond to the demand by industrialists for land, and so effectively that such demand can be steered into certain locations and also phased through time;
2. being able to determine the price of new industrial land;
3. being able to refuse to supply to certain demanders;
4. being able to impose conditions on the development, over and above the conditions possible under the planning legislation.

The additional power (item 4) to impose conditions arises because of the nature of the legislation relating to the disposal of land. If the land is sold outright, positive obligations may be and often are imposed. These include:

1. that construction must begin within a certain period and be completed by a certain date;
2. a more detailed specification of the type of industrial development than is permissible under the planning legislation. The latter regulates the physical form of development much more than the use to which it may be put.

If the land is disposed of by means of a ground lease, then even more conditions may be imposed, including such things as the obligation to maintain the building to a certain quality. It is the policy of some municipalities not to sell the ownership. A survey in 1981 of all municipalities with more than 25 000 inhabitants (this covered more than half of the national population) discovered that about 85% of municipalities disposed of land normally freehold, but 80% of municipalities used leasehold disposals as well. Thirty per cent used leasehold frequently and 12% used this exclusively (de Jonge, 1984).

An example will illustrate how far some municipalities have used their powers as suppliers of industrial land. In 1974, Rotterdam introduced a social location policy (*sociaal-vestigings-statuut*) whereby firms were allocated land only if they signed a commitment about such matters as providing accommodation for foreign workers, providing social facilities at work, and recognizing trade unions. Some other municipalities followed suit.

That this has not become more common is not because it is illegal, but because the reasons why municipalities supply industrial land have changed. The wish to steer urban development remains, of course. But from around 1980, municipalities have used the supplying of industrial land also as a means of stimulating local economic activity. The reasoning is: firms need land and will be attracted to municipalities which have a ready supply of land at low prices. As discussed above, doubt is cast on the efficiency of low land prices as a way of attracting industry, but many municipalities are still convinced that they should dispose of land as cheaply as possible, i.e. at 'cost-covering' prices (Needham, 1988).

As one might expect, the policy of trying to attract industry by offering cheap land has caused some municipalities enormous financial difficulties: they overestimated the demand, invested huge sums in land development, and could find no takers. It is to be hoped that the lessons have been learned. Another lesson which has been learned is that a municipality which combines the roles of planning authority and land supplier has an enormous power which it can arrogantly, if unconsciously, abuse. A firm which wants to build or expand contracts the municipality first in order to acquire the land, and is thereafter dependent on that municipality for building and planning permission, for some of the 'environmental' permissions, for connections to gas water and electricity, for access to the public highway, etc. Growing complaints by firms about how casually they were being treated, with long delays and excessive costs, have led many municipalities to adopt more welcoming procedures (see Needham, 1985b).

7.5.2 Local economic policy

In the Netherlands, unemployment began to shoot up around 1980, and that was the occasion for the growth of concern and action by many municipalities for the economic development of their areas – retaining existing and attracting new firms. Above, we have described how they have been providing industrial land with that aim. But they have used other means too.

Those that interest us most in the context of this chapter involve direct intervention in the market for industrial buildings: and it is characteristic of the Netherlands that municipalities have a huge involvement in the land market but avoid much involvement in the market for buildings. The measures described below are to be found in only very few municipalities:

1. the provision of work space, speculatively or to order (for some examples, see Needham, 1982, Chapter 5);

2. the provision (usually indirectly – i.e. the municipality encourages others to provide) of nursery units, to let with shared facilities such as reception, secretarial help (*bedrijfsverzamelgebouw*) (Verhoef *et al.*, 1985);
3. the provision of information about industrial space to buy or rent (*bedrijfspandenbank*; see Baetsen and Spee, 1986)

7.5.3 Regional economic policy

Throughout the 1960s and 1970s, the national government had pursued a regional economic policy aimed at reducing economic inequalities (especially in unemployment) between the regions. The main instrument used was giving grants to investment in development areas. However, in the 1980s this policy was steadily abandoned. Emphasis was placed instead on economic growth, which must take place in the regions best suited for it. The regions are now encouraged to pursue their own economic development policy, with (small) financial assistance from central government in the form of an investment grant (IPR) in a few areas in the east, north and south of the country (see also section 7.4 above). There is no restriction (apart from the normal planning and environmental controls) on economic development in the rest of the country.

7.5.4 Environmental policy

Industrial property can be affected by the environmental policy of the municipality in a number of ways. Some of these involve the issuing or withholding of the various environmental permissions described in section 7.4 above. Others involve the use of the municipality's powers as planning authority and as land supplier: those powers are, increasingly, being used to reduce the adverse environmental effects of industry. One way is by zoning within industrial estates. Environmental standards can be included in a land-use plan in so far as they relate directly to the use of land and buildings. Jurisprudence teaches that the following forms are accepted:

1. the assignment of a part of the plan area for activities which cause severe pollution or, conversely, the designation of an area where such activities are not allowed;
2. the inclusion of a list of permitted or not permitted activities (*staat van inrichtingen*);

3. the assignment of a green belt as a buffer zone for the protection of residential areas (Ashworth and Voogd, 1989, p. 29).

Another way is called comprehensive environmental zoning (*integrale milieuzonering*) and requires concerted action by the municipality and the province. There are several environmental problems, such as noise, fumes, and safety, which can be caused by industry and where the danger of nuisance decreases with distance from the source. The possibility then arises of drawing zones around large industrial establishments or estates, taking into account noise, fumes and safety all at the same time. That might seem obvious, but each type of environment problem has its own legal procedures and agencies, in addition to the procedures and agencies for land-use planning. The aim is to co-ordinate all these various procedures. So far, there is not much experience with comprehensive environmental zoning.

For some years now, the national government has wanted to reduce the use of private transport. Recently, it has announced that it wants to do this by stricter controls over the location of workplaces. Three kinds of location are distinguished: A (on a public transport node); B (on a public transport route); C (not served by public transport). Activities which attract many people (such as offices) will not be allowed on the C locations. The details have not yet been worked out, and in particular whether the existing land-use plans for industrial areas in C locations will have to be changed. If this policy is strictly enforced, it will have a significant effect on the industrial property market.

7.6 THE INDUSTRIAL PROPERTY MARKET EXPLAINED

The main distinguishing features of the industrial property market in the Netherlands are that almost all industrial buildings are built on land supplied by municipalities, and almost all industrial buildings are commissioned by the final users. In these two ways, the situation in the Netherlands is different from in most other countries. The challenge is: can those characteristics be explained?

That most industrial building takes place on land supplied by a municipality is, for the Dutch, not peculiar. It is taken for granted that most building land (also for housing, offices, shops, etc.) is provided by the municipality. The origins of that practice are to be found in the Dutch experience with reclaiming and draining land, which for technical reasons must take place on a large scale. In many parts of the country it is technically not feasible to service agricultural land for building upon a few hectares at a time. Some private development companies have, certainly in the last few years, now reached a size where they can tackle huge projects. Nevertheless, the practice, starting in Amsterdam where the private development of land in the second half of the 19th century led to unacceptable living conditions, is that land development is a task for municipalities. There are no indications that anyone wants to change this practice.

That almost all industrial buildings are commissioned by the final users is, at first sight, easy to explain. Most industrialists want space tailored to their precise requirements: such a building has a low resale value and/or is difficult to relet, and so it is uninteresting as an investment object! But that does not explain why, in some other industrialized countries, industrial buildings are an interesting investment object. It is not plausible to think that industrialists in those countries have less demanding requirements that industrialists in the Netherlands.

We suggest that the practice can be explained by the first characteristic: that most industrial land is supplied by municipalities. These see it as their task to ensure that there is never a shortage of industrial land, well serviced and with good road connections. As a result, land prices are low and stable and there is little locational variation in prices; nor is there much uncertainty, which could lead to changes in land prices (development gain). An investor in industrial property receives rents which are low and hardly rising, and also the value of his capital asset is not growing. That makes it a poor investment.

Further, acquiring land for building requires no expertise, just a visit to the town hall! Many municipalities explicitly regard the supplying of land in the same way as supplying gas, water, or electricity (Rotterdam, 1959). One of the main skills of a property developer, that of acquiring or assembling developable land, is therefore superfluous. Do-it-yourself development becomes possible and, considering the specific demands of many industrialists, attractive.

7.7 CASE STUDIES

We have seen that industrial estates are provided by municipalities, not by commercial developers, and that in most cases industrial firms build their own premises. Case studies of how that is done would give us few insights into the industrial property market. For that reason, case studies have been chosen from the few examples that there are of commercial development of industrial property as an investment good. In none of the three cases did the developer assemble the land; that was done by the municipality.

The first case, Spaarnestaete, was started in 1981 and has not been successful commercially. The second, Duivendrechtstaete, is older (started in 1975); financial information over its whole life is not available, but from 1979 at least it has been successful commercially. The third case, Alpha Business Park, has only just been completed by the developer and sold to an investor; nothing can be said about its success as an investment good, but the development process was a commercial success. It has been chosen to represent the considerable number of high-quality industrial parks at present being developed, mainly in the Randstad and, for the connections with other European countries, in North Brabant and Limburg.

7.7.1 Spaarnestaete

This complex stands on the industrial estate Waardepolder, provided by the municipality of Haarlem. It is in the north-east of that town, conveniently located for Amsterdam and Schiphol and with direct access to a motorway (but with very poor public transport). The estate has a total area of 266 ha and is now more or less full. It is a rather untidy estate (there is a car breaker's yard operating illegally opposite Spaarnestaete), but it is now improving.

The selling price per square metre of serviced land in 1980 was NFL 150, but the municipality would not sell to a property developer, only to firms building for own use. So when the developer BOZ wanted to acquire land to build this complex, it had to do so indirectly, from a private person and unserviced. This cost NFL 65/m^2; BOZ acquired 41 000 m^2 and paid another NFL 91/m^2 to get it serviced, giving total land costs of NFL 6 550 000.

The first two phases were built in 1981, when restrictions were in force on new industrial building in the prosperous parts of the country (including Haarlem), as part of regional policy. So the Ministry of Economic Affairs had first to give its permission. Phases I and II contain 26 units of equal size, each with 521 m^2 production space and 75 m^2 office space (87% to 13%); construction costs were NFL 12 305 000 and land costs of NFL 3 631 000 have been ascribed to these phases. Phase IV was built in 1986 and contains 4460 m^2 (of which 2520 m^2 is production space: 56% to 44%) divided over 16 units of three different sizes: construction costs were NFL 5 559 000 and land costs of NFL 850 000 were ascribed to it. Phase III was delayed for market reasons, and the land (11 500 m^2 and cost NFL 2 069 000) was reserved for future use. The tenant of a unit in phase I wants to expand, so it has been agreed to build for that tenant, on some of the reserve land, 860 m^2 office space and 2400 m^2 production space in four separate halls (74% to 36%). This, however, did not start until February 1991.

In 1987, the activities of BOZ were taken over by Nationale Nederlanden Vastgoed. This saw as possible tenants firms in manufacturing, construction, warehousing, repairs, transport and storage. The region in which those tenants were sought lies to the west of Amsterdam; but there, a number of similar projects have recently been completed and more are in construction.

The asking rents in 1982 for units in phases I and II were NFL 95/m^2 floorspace pa, but rents realized were somewhat lower. In 1986, average rents in the complex were NFL 100/m^2 pa; rents had not kept pace with inflation. When phase IV came onto the market in 1987, NFL 130/m^2 was asked for and realized: these latter rents are inflation-linked. Letting the units was difficult. In the first years of phases I and II, the vacancy rate was 40% to 45%; for phase IV this was 30%.

These disappointing results (high vacancies, low rent increases) led to disappointing yields. Related to historic costs (but excluding from these the ground reserved for further expansion), gross income represented a 8.46% yield and net income a 7.13% yield (in both cases, the annual average for the years 1985–89). In 1989, the complex was fully let, but there were doubts as to whether all tenants would renew their five-yearly contracts or would remove to competing complexes in the region. Furthermore, the small rent rises as a result of inflation would be eaten up by rising maintenance costs.

In 1990, the complex was sold to a Swedish investor. The price was calculated as follows. Net rental income was NFL 1 767 000 p.a. The property should give an 8% yield. Value, therefore, was NFL 22 089 000. Transaction costs (conveyancing tax and legal charges) of around 7% were to be paid by the seller. So the price was agreed of NFL 20 802 000. However, that calculation excludes the land reserved for future use. The buyer paid NFL 22 000 000 and got that land included: he acquired, therefore, for NFL 1 200 000, 11 500 m^2 of land with a book value nine years previously of NFL 2 069 000 and which had never earned a penny.

The investment rate of return calculated from 1981 for the eight years 1981–9 was 3.8%: a poor investment, caused largely by the high vacancy rate in the early years and the fact that too large a site was acquired, a quarter of which was never used!

7.7.2 Duivendrechtstaete

This complex stands on the industrial estate Weespertrekvaart, provided by the municipality of Amsterdam. Road access was poor, until a connection was made with the Amsterdam Ring Road in 1990; and public transport is still poor. The industrial estate is untidy and unattractive, but in 1990 the municipality started to improve it. The part in which Duivendrechtstaete stands is the first which is being tackled. The aim for the whole estate is to provide a high-quality industrial park with, on the land still vacant, offices, audio-visual firms, and mixed office/production spaces. Also building densities must be lower, so that vehicles can be parked on the firm's own premises.

It is the practice of the municipality of Amsterdam not to sell land but to dispose of it leasehold: prices on this estate (the price of the lump sum premium which can be paid instead of the annual ground leases) lie between NFL 37/m^2 and NFL 50/m^2 serviced. However, in this case the developer (BOZ) acquired the freehold of the plot (35 550 m^2).

The complex was built in 1975 and contains 20 300 m^2 lettable space, of which 15 800 m^2 production space and 4 500 m^2 offices (78% to 22%). this is divided between 18 units of various sizes. On the site, 200 cars can be parked. The precise land and construction costs are not known, but in 1979 the cost price (land and buildings) was booked at NFL 18 680 000.

Tenants come from the sectors storage, installations, and industrial services. Since road access has been improved, it is hoped also to attract tenants in transport and distribution. The region in which tenants are sought is the south-east of the Amsterdam area. However, there are a number of competing complexes in that region and more are being developed.

Financial details are known only from 1979, after the developer (BOZ) had become a part of Nationale Nederlanden Vastgoed. Average rents in 1981 were NFL 94/m^2 floorspace p.a. and had risen to NFL 100.50/m^2 p.a. in 1990, a rise less than inflation. In the middle of the 1980s, rents under new contracts had fallen to NFL 92/m^2 p.a., but the rents in the latest contracts are NFL 104/m^2 p.a. and are inflation-linked. Vacancy rates in the first few years are not known. However, since 1979 they have been low and little effort is needed to find new tenants. Related to historic costs, gross income represented an 8.84% yield, net income a 7.05% yield (in both cases, the annual average for the years 1985–89).

In 1989 the complex was sold to the same Swedish investor who bought Spaarestaete (see above). The price was calculated as follows. Gross rental income was NFL 2 041 000 per year, net rental income 85% of this. The property should give an 8% yield: value, therefore, NFL 21 688 000. The seller carried the transaction costs of around 7%, giving a price to the buyer of NFL 20 269 000. The actual price was a little lower: NFL 20 000 000. At that price, the seller received more than the historic cost price and the buyer an investment on which current rental income gives an initial yield of 10.2% gross, 8.7% net.

The investment rate of return calculated from 1978 for the 12 years 1978–89 was 8.4%. However, the first five years of the project are not included in this calculation. If results then were poor, that would drag down the rate calculated from the start of the project in 1975.

7.7.3 Alpha Business Park

This complex stands on the Alpha Driehoek, a high-quality industrial park provided by the municipality of Amsterdam. It has good road connections and is near a railway station (Sloterdijk). In a few years, connections will be even better, with new roads to Schiphol and to the harbour, and the coming of the high-speed tram.

The *bestemmingsplan* for this industrial area imposes strict conditions to achieve a high-quality environment. These include a *staat van inrichting* (see section 7.5 above) to exclude firms which, by causing environmental nuisance, might lower the quality. The Alpha Driehoek is divided into two parts, one for production premises, one for mixed office/production. In the former, buildings must contain a minimum of 30% office space, rising to a maximum of 70%; in the latter, buildings must contain at least 70% office space. In the whole estate, no more than 55% of a plot may be built upon; at least 20% must be used for planting; the plot ratio must not be above 1:1; the maximum building height is 19 m. The Alpha Business Park is in the part allocated for production premises.

The developer of this complex is Wilma Vastgoed, who acquired in 1986 the ground lease on 12 111 m^2 paying instead of the yearly lease over 50 years, a lump sum (premium) of NFL 375/m^2: total costs, NFL 4 500 000. Wilma Vastgoed started building in 1987, before any units had been pre-let. Construction took place in two phases, to reduce the risk; but this was an unnecessary precaution, as tenants had been found for almost all the space even before construction had been completed.

The complex contains five pavilions, two with one story, three with two. Total lettable area is 10 620 m^2, of which 4 210 m^2 are for production, and 6410 m^2 office space (40% to 60%). The production space is not only less than half of the whole, it is also suitable for light production only: ceiling height is 3.6 m instead of the usual 6.5 m, and the floor can carry a maximum 1000 kg/m^2 instead of the usual 2200 kg/m^2. Also, the space can be used very flexibly: partitions can be installed and the ceiling in the production space can be lowered with panels to convert it to office space. (The maximum permissible office space of 70% applies to the whole complex, not to individual pavilions.) On the site there are about 170 parking spaces. Construction costs were NFL 1100/m^2, total NFL 11 500 000 (excluding professional fees etc.).

Tenants were sought in the sectors computer hardware and software, electrical engineering, and distribution. The tenants were selected not only on their financial status: also they must not compete with other tenants in the complex. Moreover, the high rents should exclude unwanted tenants! Two of the five pavilions have been let to the PTT (the Dutch telecommunications utility).

In 1989, rents were NFL 175/m^2 floorspace p.a. for production space, and NFL 195/m^2 p.a. for office space. Now that distinction has been dropped, and rents for all space are NFL 195/m^2 p.a. the developer sold the complex in June 1989. The gross income at the time of the sale was NFL 1 987 000 p.a. and the developer asked NFL 26 000 000 (initial gross yield 7.6%). Some Dutch investors were interested, but only at a price which would have given an initial yield of 8 to 8.5%. A Finnish investor was prepared to accept 7.8% and paid NFL 25 500 000.

The developer had spent NFL 16m on land and construction costs, and was happy with the sale. He is now developing a similar complex (Orange Park) an an adjacent site: it will have an even higher proportion of office space.

REFERENCES AND FURTHER READING

Arthur Andersen and Co. (1991) *Investment in Netherlands Real Estate*, Amsterdam.

Alders, M. (1991) Organisational structures, in *European Environment Yearbook 1991*, Docter International UK, London.

Ashworth, G. J. and Voogd, H. (1989) Environmental policy and urban planning, in (eds) G. J. Ashworth and P. T. Kivell, *Land, water, sky; European environmental planning*, GeoPers, Groningen.

Baetsen, P. G. C. and Spee, R. P. J. M. (1986) De bedrijfspandenbank *Nijmeegse Planologische Cahiers*, nr. 23, Nijmegen.

Blankenstein-Bouwmeester, A. and Lukkes, P. (1984) Institutionele beleggers op de markt voor onroerend goed, Geografisch Instituut, RUG, Sociaal-geografische Reeks nr. 29, Groningen.

Bruil, D. W., Faludi, A. and Gastkemper, H. J. (1987) Dutch land-use laws. *Werkstukken/Planologisch en Demografisch Instituut no. 93*, Universiteit van Amsterdam, Amsterdam.

Dekker, A. (1991) Town and country planning, in *European Environmental Yearbook 1991*, Docter International UK, London.

Geus, J. W. de (1990) Kantoren, lokaties en rendement in Amsterdam, Den Haag, Rotterdam en Utrecht. Final year dissertation, vakgroep planologie, University of Nijmegen, Nijmegen.

Hogenbirk, J. and Loggers, A. (1988) Dynamiek van de commerciële vastgoedmarkt 1983–1988, doctoraal scriptie, Faculteit der Ruimtelijke Wetenschappen, RUU, Utrecht.

Jones Lang Wootton (1989/90) (1984) *The Dutch property market*, Amsterdam.

Jonge, J. de, (1984) *Gemeentelijke gronduitgifte*, Kluwer, Deventer.

Laurier, J., Graef, P. and Lapplo D. (1987) *Nieuwe patronen in de regio, Onderzoekscentrum Ruimtelijke Ontwikkeling en Volkshuisvesting*. Rijksuniversiteit Leiden, Leiden.

Lukkes, P., Krist, A. J. and van Steen, P. J. M., (1987) Kantorenmarkt, investeren en ruimte, Vonk Uitgevers, Zeist.

Needham, B. (1982) *Choosing the Right Policy Instruments*, Gower, Aldershot.

Needham, D. B. (1985a) Local government policy for industrial land in England and the Netherlands, in *Land policy: problems and alternatives* S. Barret and P. Healey, Gower, Aldershot.

Needham, B. (1985b) Verbetering van het lokale investeringsklimaat: tegan welke kosten? in *Planologische Discussiebijdragen deel III* Delftsche Uitgevers Maatschappij, Deflt, pp. 961–972.

Needham, D. B. (1988) The Netherlands, in *Land and housing policies in Europe and the USA* ed. G. Hallett, Routledge, London.

Rotterdam, Municipality of (1959) *Municipal Real Estate Policy in the Netherlands*, Rotterdam.

Rouwenhorst, D. (1991) De industrieel-onroerend-goedmarkt, final year dissertation, vakgroep planologie, University of Nijmegen, Nijmegen.

Ruimtelijke Verkenningen (1987), *Jaarboek Rijksplanologische Dienst*, Staatsuitgeverij, 's-Gravenhage.

Smidt, M. De and Wever, E, (1987) *De nederlandse industrie*, van Gorcum, Assen.

Smit, W.A.P. (1989) *Huurrecht bedrijfsruimten*, 3rd edn, Kluwer, Deventer.

Spit, T. (1987) Town and country planning, in *European Environmental Yearbook 1987*, Docter International UK, London.

Statistiek van de investeringen in vaste activa in de nijverheid (1988) Centraal Bureau van de Statistiek, Voorburg.

Vaart, D. van der and Weesep, J. van (eds) (1987) *Stedelijke Netwerken: voorstudies*, Ministerie Onderwijs en Wetenschappen, Amsterdam, Delft, Utrecht, 1987.

VastGoedMarkt (1989) Katern 2: Bedrijfsterreinen, vol. 16 no. 11, 3 November 1989.

Verhagen, J. (1989) Flexibiliteit in the gemeentelijke ruimtelijke-ordenings-praktijk, in (eds), *Ruimtelijk handelen* eds B. Needham and N. Muller, Kerckebosch, Zeist.

Verhoef, L. H. J., Haar, H. and Ewolds, K. H. (1985) Het functioneren van bedrijvencentra: een tussenbalans, in *Kroniek van het Ambacht/ Klein- en Middenbedrijf*, pp. 21–4.

VROM (1990) Leegstand en hergebruik, een nieuwe market voor de bouw, Ministerie Volkshuisvesting, Ruimtelijke Ordening, Milieu, 's-Gravenhage.

Waveren, P. C. van (1990) Bouw bedrijfsruimten toonde sterke fluctuaties in de jaren tachtig, in *VastGoedMarkt*, December.

Zadelhoff Makelaars (1990) Visie achter de feiten 1989–1990, Zadelhoff Makelaars, Amsterdam 1990.

8

SWEDEN

Hans Mattsson

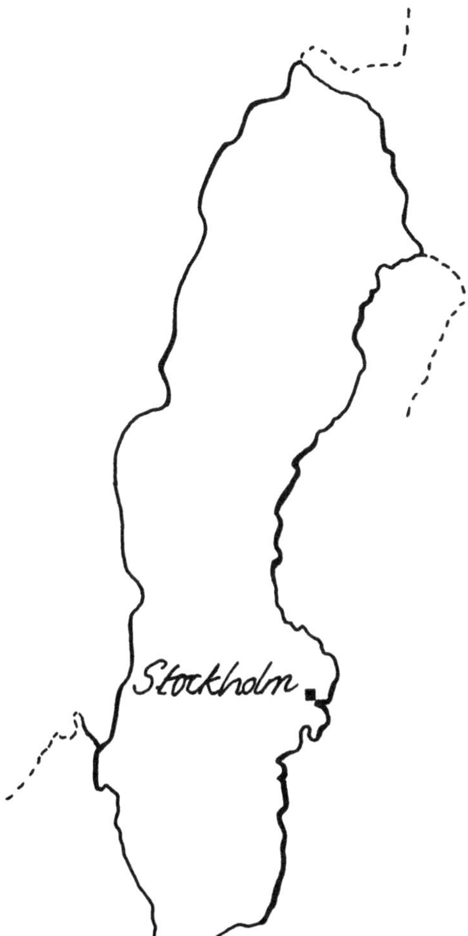

Sweden, with its 450 000 km^2, is barely the size of France, a bit bigger than Germany and almost twice as big as the United Kingdom. Half its area is productive forest land, just over 5% is agricultural land and 1–2% is urban land. The rest is made up of lakes and watercourses, mountains and other impediments. Sweden's 8.5 million inhabitants live mainly in the southern parts of the country and along the Gulf of Bothnia coast. The country is divided into 284 municipalities. The largest urban regions are Stockholm (1.5 million residents), Gothenburg (700 000) and Malmö (500 000). Another eight municipalities have more than 100 000 residents each. Sweden's population density is 20/km^2. The most densely populated region is Stockholm, with 235 residents/km^2, and the most sparsely populated is Northern Norrland, in the far north of the country, with 3 inhabitants/km^2.

The active labour force numbers 4.4 million people. About 20% of them are employed in manufacturing, nearly half in 'metal bashing' and about one-fifth in wood conversion and papermaking. Another 300 000 are employed in the construction industry. There are upwards of 20 000 manufacturing firms with employees, including 1200 whose personnel strength exceeds 100. Manufacturing concerns are located all over the country (SCB, 1989a).

8.1 PLANNING FOR INDUSTRIAL LAND AND PROPERTY

8.1.1 The formal planning system

In principle, landowners in Sweden are only entitled to maintain the current land use. A new use has to be officially sanctioned. If a significant restriction is imposed on the right of current land use, the landowner will normally receive compensation. However, any appreciation resulting from a change in land use accrues to the landowner. Official decisions concerning changes in land use can be based on physical planning and are frequently made by elected bodies. This is particularly the case with decisions on new development, for which the politically governed municipality (*kommun*) is often the sole decision-making body. The municipality, however, has to defer to current legislation and also to other authorities, such as the government, the county administrative board (*länsstyrelse*), the National Franchise Board for Environment Protection (*koncessionsnämnd för miljöskydd*) and the property subdivision authority (*fastighetsbildningsmyndighet*).

Statutory provisions on physical planning are common to all urban planning, but there are various supplementary provisions relating to industrial development. Most of the planning provisions are to be found in the Natural Resources Act (SFS 1987: 12) and the Planning and Building Act (SFS 1987: 10), both of which will be described in general terms together with various other enactments.

Most of the provisions of the Natural Resources Act (NRA) cannot be applied independently, but must be co-ordinated with official decisions under other legislation. The NRA stipulates that the physical environment must be used in a manner conducive to good long-term management in the ecological, social and economic senses. Land and water areas must be applied to the purpose for which they are best suited in view of their nature, their location and existing needs. In large areas of Sweden, new development and other interference with the environment is forbidden if it would be detrimental to the natural environment and the cultural heritage. This applies, for example, to much of the coastline and, inland, to principal waterways. This provision, however, does not stand in the way of the development of existing urban communities or of local enterprise within the areas concerned.

More detailed provisions on planning and building are laid down in the Planning and Building Act (PBA), the aim of which is to promote urban development with good living conditions for all, while at the same time showing consideration for personal liberty. Land and water areas must be applied to the purpose for which they are best suited. The Act, however, does not define 'good environments' or 'suitable use'; instead it leaves these matters to be decided by the municipalities. This is because the PBA makes the planning of land and water areas a municipal concern, and at the discretion of the municipal council to decide where, when and how plans are to be adopted. The municipality is also expected to play a very active part in the drawing up of land-use plans. The Government and Swedish parliament (*Riksdag*), however, have reserved certain powers of planning and control in areas of national interest and also for the protection of individual health and the prevention of accidents.

Several planning instruments are defined in PBA, including the regional plan, the comprehensive plan, special area regulations, the detailed plan and the subdivision or parcelling plan. The regional plan (*regionplan*) covers several municipalities and very limited use is likely to be made of it in future. The comprehensive plan (*översiktsplan*) roughly outlines land use for the entire municipality and is intended to summarize and illustrate municipal assessments of the future. An accompanying commentary can indicate long-term municipal aims, e.g. for enterprise development. If anything, the comprehensive plan resembles a UK local plan at district level. These comprehensive plans are not legally binding; they are merely for the guidance of decision-makers. Special area regulations (*områdesbestämmelser*) can supplement the comprehensive plan by defining, with legal effect, certain basic features of land use in a limited area of the municipality. For example, they can define safety zones surrounding industrial areas.

The detailed plan (*detaljplan*) can regulate land use in great detail for limited areas of the municipality. Detailed plans are legally binding documents which have to be complied with by the municipality, landowners and others wishing to build on a property or otherwise alter its use. An adoptive detailed plan entitles the landowner to change to a specified form of land use within a limited period which the plan also specifies. The subdivision plan (*fastighetsplan*) regulates future land parcelling in detail. This type of plan can only be used within the detailed plan. Finally, building requires a building permit (*bygglov*). A permit can also be applied for land for which no plan has been prepared, but it is uncommon for permission to be given for industrial development before a plan has been drawn up for the actual land.

The content of the detailed plan, subdivision plan and building permit, and the links between them, demand closer description. The function of the detailed plan is to test the suitability of the land for development and to regulate the design of the developed environment. The plan has to indicate and demarcate public spaces such as roads, streets, squares and parks and also development sites. The mapping scale employed is usually 1:1000 or 1:2000, so that areas destined for different uses can be distinctly demarcated. The purpose of the development site has to be determined. It can be referred to, for example, as 'industrial land', 'light industry and offices' etc. Only the activity specified may be established within the area. The rider 'the activities may not have a disruptive effect on nearby housing development' imposes further restriction on use.

The detailed plan may also indicate the maximum extent of building above and below ground, as well as the use, positioning and design of buildings. Installations and the level of the ground can be indicated. Provisions can be included on fencing, exits into public spaces and the positioning and design of parking facilities. The plan can also indicate the maximum permissible levels of disturbances such as air pollution, noise, vibrations and light. The principles of property subdivision, e.g. maximum and minimum plot sizes, can be determined.

A detailed planning proposal cannot be adopted until those whose interests are affected have been consulted. So it is up to the property owners and other interested parties to play an active part in the consulting process and also during the planning process generally, so as to make their views felt. After this consultation (which often takes the form of a public meeting), the planning proposal also has to be put on public exhibition for at least three weeks. After the exhibition, the proposal can be adopted by the municipal council. Those whose interests are affected can lodge a limited appeal, first with the county administrative board and then with the Government.

As stated earlier, the detailed plan is a legally binding document. Once it has been adopted and the adoption has acquired force of law, the landowner is entitled, within the time-limit set for implementation, to develop the land in accordance with the plan, but not at variance with it. The length of the implementation period, decided when the plan is adopted, must be between five and fifteen years. The period can then be prolonged for five years at a time. After the implementation period has expired, the municipality may amend or repeal the plan without having to compensate property owners for loss of development rights. The municipality may also acquire development sites which have not been developed essentially in accordance with the plan.

Detailed plans have to be up to date and feasible. They may not cover unjustifiably large areas in relation to what can be accomplished during the implementation period. The role of implementation is underlined by an implementation schedule (*genomförandebeskrivning*) which the plan includes. That description is a special document which has to present the organizational, technical, economic and legal measures necessary for a co-ordinated, practical implementation of the plan.

The highly detailed nature of detailed plans means that they are very liable to become out of date. Besides, it is unreasonable to expect the authors of a plan at all times to predict the changing needs of the market in a dynamic society. Extensions and new buildings in an area for which a plan has recently been adopted may demand a planning amendment, and all parties may be agreed on this. What in practice is a protracted business of planning work, consultation, exhibition and manifold municipal decision-making has to be started all over again, perhaps merely in order to revise a minor detail. A simplified plan-approval procedure (*enkelt planförfarande*) is frequently employed to make things easier in this kind of situation. This method presupposes absolute consensus, in which case consultation and other procedures can be simplified.

An adopted detailed plan affects a property owner in a number of ways, and not only in terms of development rights. The municipality is normally duty-bound to build streets and to take charge of their maintenance. Construction and improvements can be charged to the account of the property owners, while maintenance is paid for out of taxation revenue. Normally the municipality will also be responsible for water supply and sewerage networks in urban areas and can charge the property owner for both construction and operation (SFS 1970: 244). In addition, the municipality becomes entitled to acquire land for communal purposes.

A detailed plan is sometimes prepared for an area where property subdivisions are discontinuous or otherwise inappropriate. As a convenient way of achieving a practical subdivision of properties or otherwise facilitating the implementation of a plan, a subdivision plan can be adopted for the development sites of the detailed plan. The preparation procedure is the same as for a detailed plan, and both types of plan can be drawn up concurrently. The property subdivision plan shows future plots with exact boundary points marked to the nearest centimetre on the planning map. If a planned property comprises land belonging to several different properties, each of the landowners may request that the land be transferred to him for the formation of the new property. Compensation is paid to those who have to relinquish land.

When a detailed plan and a subdivision plan, if any, are adopted, land parcelling can take place if necessary. New properties are normally formed by subdividing an old one. Part of one property may be transferred to another by re-allotment. All property formation has to comply with current plans (SFS 1970: 988).

This is a suitable stage at which to apply for a building permit. The application must be allowed if the proposed measure is not at variance with current planning regulations. In addition, the property on which the measure is to be undertaken must tally with the detailed plan and also with the subdivision plan if there is one. All prospective developers – State and municipality included – have to apply for building permits. Permits are awarded by the municipal building committee (*byggnadsnamnol*). Exceptions notwithstanding, the main principle is that permits have to be obtained both for new buildings and for extensions and alterations. Permits are also required when existing buildings are to be partly or wholly diverted to a new purpose, and also for alterations to vital parts of existing buildings, e.g. load-carrying structures. Changes in the colour of a building and changes of cladding and roofing, as well as the erection of signs and lighting arrangements, are normally subject to the award of a permit within the detailed plan for industrial development. Building permits are also needed for the construction of facilities. This applies to the arrangement of depots and yards, the erection of walls and wooden fences, and the provision of outdoor parking spaces. Within a detailed plan, a demolition permit (*rivningslov*) must be obtained for demolishing buildings or parts of the same, unless otherwise provided in the plan.

Buildings are subject to a number of technical requirements. They must be compatible with sound energy management. Their basic structure, carcases and other load-carrying parts must be securely designed. They must be suitable for their purpose and compatible with comfort, good hygiene, a good working environment and a satisfactory indoor climate. They must afford protection from accidents and fire. Waste must be properly removable. The Swedish National Board of Physical Planning, Building and Housing (*Boverket*) issues regulations and general recommendations (BFS 1988: 18) on the implementation of these provisions. Building permits for working facilities for employees may only be granted if the Labour Inspectorate (*yrkesinspektionen*) has commented on the suitability of the measure intended and if the statement thus returned makes it clear that a safety delegate, a safety committee or an organization representing the employees has had an opportunity of commenting on the application.

In addition to a building permit, another type of permit, called *byggnadstillstånd*, may be necessary for projects exceeding a certain level of expenditure (SFS 1971: 1204). These permits are issued by the county labour board (*länsarbetsnämnden*), a national authority, which scrutinizes the building scheme in relation to the employment situation within the construction industry.

Planning work up to the adoption point probably takes two or three years to complete in normal circumstances. Actual periods can be either longer or shorter. The final building process moves a good deal faster as a rule.

8.1.2 Environmentally hazardous activities

Persons conducting environmentally hazardous activities can come up against the Environment Protection Act (SFS 1969: 387), which is concerned with pollution and other environmental disturbances caused by permanent installations. This Act applies to emissions of waste water, solid substances or gas from land, buildings or facilities, and also to more permanent disturbances caused by air pollution, noise, vibration, light or suchlike.

The National Franchise Board for Environmental Protection (NFB

Koncessionsnamnden for miljoskydd) can grant permission for such activities in response to an application by the person conducting them or intending to do so. Certain types of industry and other installations may not even be established or modified without a licence, as is also the case with discharges of certain effluents. The Environment Protection Ordinance (SFS 1981: 574) includes a list of activities for which permits are required. The application procedure requires the applicant company to give a very detailed account of the consequences of the project and of future control programmes. If it sanctions the activities, the NFB must indicate the type of environmentally hazardous activity referred to and the conditions which are to apply. A permit may be awarded for a limited period if there are special grounds for doing so. The person or company conducting the activities must bear the expense which the NFB's restrictions may entail, e.g. costly purification equipment.

An activity cannot be prohibited once a concession has been awarded for it. However, the person conducting the activity may in certain cases be ordered by a court to pay damages to neighbours and others suffering damage or inconvenience as a result of the activity (SFS 1986: 227).

8.1.3 Other stages in the development process

Municipalities often pursue an active land policy in planning areas before or during the planning process. This involves them in purchasing and, sometimes, expropriating land. The right of expropriation applies to land which is to be used for urban development. Purchase and expropriation may be intended, through municipal title to the land, to facilitate subsequent planning implementation. Expropriation is uncommon, but it serves as a means of bringing pressure to bear on land purchase negotiations.

If the proposal concerns private land, a development agreement will often be drawn up between the municipality and major landowners. That agreement can deal with matters of liability and time-scales for building development and the construction of streets, water mains and sewerage. The agreement may require the property owner to build municipal streets and water and sewerage facilities, in return for which he will not be charged for them subsequently. Additional questions which can be raised are the financing of municipally constructed facilities, the conveyance of land for public spaces to the municipality, property subdivision and easements. The development agreement may define penalties in the event of a party's not discharging his obligations, and may specify the economic security which the developer is to furnish for his commitment. In this way, the municipality can fulfil the developer's obligations if he himself defaults on them.

A municipality wishing to draw up an agreement can often negotiate from a position of strength, because it can refuse to adopt the detailed plan until the development agreement has been signed. The municipality is not entitled to secure economic advantages for itself over and above the charges it can require the property owners to pay for streets, water and sewerage and planning. The Swedish law of contract imposes restrictions in this regard.

If the municipality owns the land, which is not uncommon in the case of new industrial estates, it can subdivide properties and convey the plots to various interests. Sometimes conveyance is combined with a land allocation agreement which, by and large, can contain the same provisions as the development agreement mentioned earlier. Economically, however, this puts the municipality in a better position, enabling it, as the seller of the land, to cash in on planning benefits. It can do so by exacting payment but also by imposing obligations on the developer.

The building process involves a number of organizationally complicated stages which will not be dealt with here. Suffice it to say that developer and contractor often make use of standardized contracts and job descriptions (AB72, ABT74 and AMA) for the transactions involved.

8.2 PROPERTY RIGHTS

8.2.1 Real estate

Swedish law distinguishes between real estate (*fast egendom*) and personal property or chattels (*lös egendom*). Real estate means land divided into property units. All other property is personal property. In many respects, these two kinds of property come under completely different legal rules. Our attention here will be confined to the rules governing real estate, unless otherwise indicated.

The main enactment dealing with real estate is the Property Code (SFS 1971: 1209), which lays down: 'Real estate is land. This is divided into properties . . .' Properties, then, are areas staked out on the ground. Theoretically they include everything below ground and the air space above the ground as well. There are, however, exceptions to this rule. For example, mining rights, in principle, are not included. Properties cannot be stratified in three dimensions. Different storeys of a building cannot constitute separate properties with different owners; nor is it possible for the area below ground to have one owner and the ground surface another.

The individual property has appurtenances which are unconditionally attached to it. These comprise a building, a conduit, fencing and other fixed installations in or above ground for permanent use. The building includes fixtures and other items intended for permanent use such as a lift, a water main, electric points and a central heating boiler. A property partly or wholly equipped for industrial activity normally also includes machinery and other equipment added to it mainly for use in the activities conducted on the property. If, however, the owner makes and publishes a special declaration to the contrary, equipment of this kind need not belong to the property. Vehicles, office equipment and hand tools can never do so.

All land, in principle, is divided up into properties, each of them registered under a unique designation (*fastighetsbeteckning*). Every property has an owner. Ownership and its content, however, are not defined by law; instead, the official doctrine has it that an owner's title includes everything of which the law has not deprived it. There is, for example, no general right of building, because building permits have to be applied for and applications can be refused.

Property subdivisions may only be changed by specially authorized land surveyors (*lantmatare*) employed by national and, in certain cases, municipal subdivision authorities. These surveyors can make decisions on their own authority. Changes can mean the formation of new properties or the re-allocation of old ones. A new property can be created by detaching an area (*avstyckning*) from a pre-existing property (SFS 1970: 988). The new parcel can then be bought and sold separately from the original property. A property with several joint owners having definite shares in it can in certain cases be partitioned (*klyvning*) so as to make a separate property of each part-owner's share. The owner of two or more properties can amalgamate (*sammanläggning*) them into a single property. If property owners want an area of land to be transferred from one property to another, this can be done by re-allotment (*fastighetsreglering*). A cadastral map is drawn up for every change of property subdivision. In all subdivision of property, the surveyor must investigate whether the new properties are enduringly suitable for their purpose and whether they conform to any detailed plan or subdivision plan for the area. If the surveyor finds that the subdivision application cannot be put into effect, the transaction has to be called off. All decisions can be contested by appeal.

8.2.2 Registration of real estate

Real estate forms the subject of a co-ordinated, computerized title and property register (*inskrivnings- och fastighetsregister*) which is planned to include all properties by the mid–1990s. In the meantime manual registers of similar content are being maintained in parts of the country. These registers are in the public domain and therefore of particular value to persons requiring an instant view of the legal conditions relating to a property. A brief description will be given of the computerized register.

The locations of properties are shown by separate register maps, scaled 1:2000–1:10 000. The register gives the name and number of each property and its ownership. It also shows the date on which title deeds were registered (*lagfart*), i.e. when the ownership of the property was entered in the register. Frequently it also gives the purchase price of the property. If a new owner of the property has applied for registration of title deeds but, for some formal reason, this has not been granted (dormant registration), this is made clear. Mortgages on the property are entered at their face value, often together with the name of the mortgagee. The register also gives particulars concerning mortgaged easements and usufructs charged to the property. 'Official easements' formed by legal survey procedure, may be shown. Any distraint, bankruptcy, executive auction or expropriation is noted. The register contains particulars concerning the area of the property and the legal survey procedures completed for it. The existence of legally valid plans is shown, as well as certain other provisions concerning land. The assessed value of the property is also shown.

8.2.3 Purchase and sale of real estate

Real estate can be freely bought and sold, with certain exceptions, but transactions are only valid if certain formalities are complied with.

1. A written document of purchase has to be drawn up. This document must include a declaration that the seller conveys the property to the buyer.
2. The purchase price has to be stated. Transactions not meeting these formal requirements are invalid. A binding agreement exists when both parties have signed the document. If only one party has signed, the agreement is not binding on either party. It is important to note that the Swedish Property Code does not accept options in the form

of a promise concerning the purchase or sale of real estate. Although only one purchase document is stipulated, the usual practice is to draw up this document in duplicate. First the parties usually draw up an exhaustive purchase agreement containing all the provisions agreed on, e.g. purchase price, payment date and possession date. When the purchase price is paid, a deed of purchase is made out.

A person who has acquired a property must apply to the registration authority to have the acquisition (the title deeds) registered. The title deed registration application must specify the acquisition document. If an acquisition permit is required from another authority, this too must be appended. It is important to note that aliens, foreign companies etc. normally have to obtain acquisition permits in order to purchase real estate for commercial purposes. A title deed gives publicity to the purchase transaction. After this the seller can no longer dispose of the property improperly. For example, he cannot mortgage it or otherwise convey a special title in it.

The municipality has a right of pre-emption over certain properties, and the title deed application date is normally the starting point for this three-month option (SFS 1967: 868). The right of pre-emption means that the municipality may acquire the property sold to another purchaser, on the terms which that purchaser and the seller have agreed on.

As stated earlier, aliens, foreign companies etc. may not acquire real estate or even site leaseholds in Sweden without an acquisition permit (SFS 1982: 618). There are minor exceptions to this rule. Acquisition permit questions are normally dealt with by the county administrative board, but in some cases they are referred to the Government. An acquisition permit must be granted if there is no objection relating to the suitability of the property for the intended purpose or for the circumstances of the person acquiring the property. Acquisition permits can be refused with reference to national defence or other public interests. Acquisition permits are also needed for the purchase of shares in Swedish companies, and for holdings in trading partnerships or other businesses operated in Sweden (SFS 1982: 617). In this way too, consequently, it is impossible to acquire real estate without a permit application procedure. Applications are made to the Government in the case of companies with more than 500 employees, and otherwise to the county administrative board.

So far we have been dealing with purchases of entire properties. It is also possible to buy shares in properties or to purchase an area of land not constituting a property unit. When shares are purchased (e.g. in the case of two or more persons jointly acquiring a property), the joint owners will have intangible shares in the property and neither of them will own an area distinguishable on the ground. If the joint owners have not entered into any joint ownership agreement when acquiring the property, they must always agree between themselves on any measures to be taken on the property, otherwise no such measures can be taken at all (SFS 1904: 48).

If an area of land is purchased which does not constitute a property in its own right, application for subdivision of the area (or else for re-allotment of the land area from the seller's property to the buyer's) must be addressed to the property subdivision authority (*fastighetsbildningsmyndighet*) not more than six months after the deed of purchase was drawn up. Failing this, or if the property subdivision is not feasible, the purchase will lapse. Current planning regulations etc. may oblige the property subdivision authority (the surveyor) to reject the application.

8.2.4 Mortgage deeds and advancements of credits

A brief account will be given of real estate mortgage deeds (SFS 1971: 1209). Mortgage deeds are issued by the property registration authority at the property owner's request. The date of issue and the exact amount in Swedish kronor (SEK) are entered in the property register and in the mortgage deed which is issued simultaneously. The mortgage deed belongs to the property owner. The property owner can then transfer the mortgage deed as security for a debt. More detailed terms are recorded in a special instrument of debt drawn up by the property owner and the creditor. This instrument of debt shows payment periods, rate of interest and other terms. Delivery of the mortgage deed completes the mortgage transaction. The mortgage deed is the creditor's security for a loan advance, with the ultimate recourse of distraint. The order of priority between two or more mortgage deeds is based on the chronological order in which they were issued. The first deed to be issued carries the best security.

The mortgage deed can also be pledged to a creditor before the property registration authority has even consented to the mortgage and issued the deed. This often happens when loans are contracted with a credit institute. In a case of this kind the mortgage deed is delivered straight to the mortgages and not to the property owner.

Personal property used in commercial activities can also be pledged to secure credits by means of chattel mortgages (*företagshypotek*, SFS 1984: 649). The entrepreneur applies for a certain amount to be charged to the business. This charge is entered in a central computerized register for the whole of Sweden. After obtaining the chattel mortgage deed, the entrepreneur can assign it to the creditor as security for the debt. With certain restrictions, the chattel mortgage includes all personal property in the business at any given time.

8.2.5 Site leasehold, facility lease and rental tenure

The property owner can transfer an easement and usufruct in his property to another person through an agreement under civil law (SFS 1971: 1209). An easement causes a property to carry the right of using part of another property for a certain purpose, e.g. an access road, for an indefinite period of time. An easement agreed on in writing may be entered in the title deed register to protect the proprietor's rights when the subservient tenement changes hands. One has to distinguish between this contracted easement (an easement by agreement) and an official easement. The latter results from an official decision, e.g. legal survey procedure, after which it cannot be eliminated without renewed official examination.

Usufruct may refer to site leasehold (*tomträtt*), various other forms of leasehold (*arrende*), tenancy (*hyra*) and other rights of use (*annan nyttjanderätt*). These rights are conveyed to a particular person, company etc. Three types of use will be described in somewhat greater detail here, because they may be of interest to industrial entrepreneurs. These three types are site leasehold, facility lease (*anläggningsarrende*) and tenancy.

Site leasehold is a form of conveyance whose legal consequences, from the viewpoint of the site lessee, resemble those of freehold tenure. For example, the site lessee can apply for registration of his title in order to obtain a mortgage deed. Site leasehold may only be conveyed for publicly owned property and is principally intended to enable the community (the State and municipalities) to benefit from the appreciation of land values, while entitling the site lessee to enjoy the property. Site leasehold is conveyed for an indefinite period and for a particular purpose, in return for an annual ground rent or site leasehold charge paid in cash. The purpose of site leasehold has to be stated in the conveyance agreement and can take the form of industrial activity. The site lessee is fully entitled to convey the site leasehold to another person. Ground rent remains unchanged for certain periods of time, the minimum period being 10 years, and it has to be fixed with reference to the value of the land for the intended purpose. (The regulations are being reviewed at the time of writing.)

The site leasehold agreement may not be cancelled by the site lessee. The property owner, however, may cancel it after predefined periods. For industrial purposes, these periods may not be less than 20 years. Cancellation is only possible if it is important to the owner to be able to apply the property to a different use. On the termination of the site leasehold, the property holder must purchase buildings and other appurtenances of the site leasehold. For industrial properties, however, agreements may be concluded to the effect that only a certain measure of compensation, or none at all, is to be paid.

The title deed register mentioned earlier includes particulars concerning site leasehold: the purpose of the conveyance, the ground rent, readjustment periods, any cancellation of the agreement, the name of the site lessee, the date of taking possession, and the purchase price when the site leasehold was acquired from a previous proprietor. Registration of a site leasehold corresponds to the issue of a title deed.

A facility lease is a form of conveyance of land in return for payment. What happens is that a lessee becomes entitled to erect and preserve a building on the leased property for commercial purposes. The lessee cannot apply for registration with a view to obtaining a mortgage deed, because buildings on leased land rank as personal property. Companies which have erected industrial buildings on lease land, therefore, must have recourse to chattel mortgages.

Leasehold agreements have to be drawn up in writing and must contain all the conditions of the leasehold. The lessee is entitled to register his title in order to protect it against a new property owner, unless otherwise agreed. The agreement has to be concluded for a specified period. This must not be less than one year, but an agreement cannot be binding for more than 25 years within a detailed plan and 50 years outside it.

Industrial concerns often rent facilities, and there is a great deal of discretion where tenancy agreements are concerned. The following account gives the main principles involved. A tenancy agreement has to be drawn up in writing if the landlord or tenant so require. A written agreement may be registered to protect the tenant's title, unless otherwise agreed. However, the tenant cannot request title deed registration for the issue of a mortgage deed. The tenancy agreement lays down what the facility is to be used for; who is to maintain it, the duration of the tenancy, the rent payable and the items, respectively, included and not included in the rent (heating, water and sewerage, refuse collection, snow clearance etc.).

Tenancy agreements for facilities (non-housing) usually run for three years with a nine-month period of notice. The rent is index-adjusted during that period. Index adjustment presupposes a fixed rental period of at least three years. The tenacy agreement has to be cancelled on the expiry of the agreed term, unless the agreement includes a renewal clause. If the agreement is not cancelled, it will remain in force but without index adjustment.

The tenant has indirect security of tenure when the tenancy is cancelled, because the law is constructed on the assumption of the landlord – tenant relationship normally continuing. One important basic principle is that in certain cases the tenant is entitled to compensation for relocating. When cancelling the agreement, the landlord must state his conditions (e.g. a new rent) for renewing the agreement or his reason for refusing to renew it. If the tenant does not agree to the landlord's demands and also refuses to move without compensation, then the tenant must refer the dispute to the regional rent tribunal (*hyresnämnden*). The regional rent tribunal will then propose a new rent, adjusted to market conditions. If the parties are unable to agree on this the tenant has to move. If the landlord is then unable to let the premises at a substantially higher rent than that proposed by the regional rent tribunal, the old tenant is entitled to compensation.

The landlord can refuse to extend the agreement if the building is to be demolished or extensively altered. In this case the landlord has to provide acceptable replacement premises, otherwise the tenant is entitled to compensation and to reimbursement for his removal costs. The tenant can also cancel the agreement in order to get it amended or to move out. The same periods of notice apply to the tenant as to the landlord. The tenant may not transfer his tenancy without the landlord's consent. If the tenancy is transferred in connection with the sale of the entire business, then it may be transferred with the permission of the regional rent tribunal.

8.2.6 Sale and lease-back

Sale and lease-back is a new phenomenon in the Swedish property market. What happens is that a property owner, A, sells his property to a property company, B. An ordinary purchase agreement is used for this. At the same time A rents the property back from B. A tenancy agreement is drawn up. Finally the property company, B, issues a special pledge (option), promising to sell the property back to A again at a certain price after a number of years.

The option agreement confirms the tenant's right to cancel the tenancy agreement and re-acquire the property at certain points in time. Option agreements providing for the re-purchase of real estate, however, are invalid, and so agreements of this kind are really based on trusting the property company to actually sell the property. For this reason, the sale and lease-back market has instead come to be based on the transactions proceeding through the sale and optional re-purchase of trading partnerships, limited partnerships or joint stock companies, which in turn are the owners of the properties. The option is then valid because it refers to personal property, i.e. holdings in companies. (The regulations for sale and leaseback are currently being reviewed.)

8.2.7 Taxation

A word should also be said about taxes. When title deed registration is granted, stamp duty becomes payable. This is 3% of the purchase price if the purchaser is a legal corporation, and 1.5% otherwise. The stamp duty for mortgage deeds is 2% of the mortgage value.

Real property is value-assessed and its assessed value forms a basis for computing the property tax payable by property owners. From 1991 onwards, the property tax on commercial properties will be 3.5% of assessed value. There is, however, no property tax on industrial properties. Capital gains tax will have to be paid on any sale of the property. To this end the nominal profit on the sale (sale price minus purchase price) will be taxed at 30%. In certain situations it will also be possible to deduct repairs from the profit.

8.3 THE PROPERTY MARKET

8.3.1 Industrial units

It is easier to describe the code of rules for real estate than to describe the property market. The code of rules is reasonably stable over time, while the market varies continuously with supply and demand. What is more, the remarks offered below are to a great extent based on personal contacts with people in the property trade and not on systematic surveys, with all the uncertainty this implies. All monetary figures refer to Swedish kronor (SEK). One kronor is equivalent to 0.13 ecu.

An overview of the stock of industrial properties can be obtained from published digests of the nationwide property assessments. Real estate and buildings on land belonging to others are value-assessed at regular intervals in Sweden. The latest assessments of industrial units took place in 1981 and 1988, and the next is planned for 1994. If extensive changes are made to a property between these points in time, a special assessment is made. The assessed value always refers to market value at a particular point in time and is 75% of estimated market value. The assessment level for the 1988 assessment and for subsequent special assessments until 1994 is to be 75% of market value in 1986. Every new property assessment is preceded by extensive test valuations of large property stocks, aimed at achieving reasonably well-founded market values for different types of property. Purchase prices, rents, buildings and facility production costs etc. are all analysed.

Assessment is usually based on real estate units. But an assessment unit can also comprise several properties, as well as a functional part of one property. But even though assessment units sometimes differ from the property unit concept, the assessment statistics convey a good picture of the number of industrial properties. There are more than 10 000 development sites (Table 8.1). These are sites which can be immediately applied to the purpose for which they are intended. There are nearly 1000 undeveloped land units, and the number or renewal sites is 500. Undeveloped land comprises units requiring little preparation before building development, and renewal units comprise land with buildings scheduled for demolition. The three groups of land can be looked on as the immediate land potential for new industrial development. Municipal planning preparedness for providing new industrial land, however, is frequently high in municipalities looking for industrial start-ups. In such communities, additional land can be obtainable at relatively short notice.

There are 43 000 industrial units, including warehousing etc. The large number of units for the timber goods industry (4700) testifies to the national importance of forest industry. About 24 000 units are used by manufacturing industry (the first six groups under the heading Industry units and warehousing). Their combined assessed value is SEK 100 000m. Statistics Sweden makes continuous comparisons between purchase prices and assessed values in these groups (SCB 1989c and 1990b). Those statistics show that selling prices for 1989 were on average about 100% greater than assessed values. The combined value of industrial properties, therefore, should be upwards of SEK 200 000m. (It is uncertain whether the other assessment units can be written up by the same percentage.) There are just over 1000 advance factories (*industrihotell*) – that is, industrial buildings erected for industrial enterprises to rent. There are also about 12 600 warehousings, 9000 petrol stations and repair shops, 1400 electricity production units (mostly for

Table 8.1 Industrial units in 1989. Number of tax-assessed units and total assessed values (SEK millions)

	Assessed units	Assessed value
Industrial land		
Undeveloped land	903	504
Development sites	10 641	4 177
Renewal sites	517	163
	12 061	4 814
Industrial units and warehousing		
Chemical industry	709	14 354
Food industry	1 591	8 391
Metal manufacturing and machinery industry	6 543	27 527
Textile and garment industry	832	1 509
Timber goods industry	4 672	14 059
Other manufacturing	9 959	43 172
Advance factories	1 010	5 840
Industrial units with other buildings	4 680	5 992
Warehousing	12 568	15 369
Depots yards	840	187
	43 404	136 400
Other units		
Petrol stations	2 954	2 602
Repair shop	6 039	5 245
Power production units	1 372	127 525
Quarrying etc	4 606	1 007
	14 971	136 379

(Source: SCB, 1990a)

hydro power) with a very high combined property value, and 4600 units for quarrying and similar activities. In addition to the figures shown in the table, there are 1150 separate office units on industrial land, directly adjoining industrial premises. For 1988 there is a division of industrial

Table 8.2 Various industrial units in 1988, by assessed value

Assessed value (SEKm)	Metal manufacturing (%) etc.	Timber goods industry	Other manufacturing (%)	Warehousing (%)	Advance factories (%)
0.0– 0.2	21	43	25	55	2
>0.2– 1.0	35	31	35	26	20
>1.0–10.0	38	23	34	17	65
>10.0	6	3	6	2	13
Total	100	100	100	100	100

(Source: SCB, 1989b)

units by assessed values. A selection of the numerically dominant industrial categories will be found in Table 8.2. Advance factories are also included, because they are interesting from a management viewpoint.

As can be seen from the table, the property stock is heterogeneous, including many small units and occasional large complexes. Depending on the type of industrial facility, between three- and four-fifths of the assessment values in these groups falls short of SEK 1m. Even really low values (under SEK 200 000) are common, suggesting that there are plenty of small, simple units of low value.

The advance factories deviate from this pattern. Only about 2% of them have low assessed values, while three-quarters are assessed at more than SEK 1m. A quarter (23%) of all advance factories are located in the Gothenburg and Malmö regions, whereas these regions have 14% of Sweden's population. In the Stockholm region, on the other hand, the two percentages are roughly equal (16% and 18% respectively). This latter fact may seem remarkable, since turnover in the office rental market shows that big-city companies are often very mobile. Industrial concerns should be very interested in renting, so as to facilitate changes in response to current facility requirements. Small and new companies in particular need to rent premises, lacking as they do the resources for real estate investment. The reason for the low percentage in the Stockholm region may be that the region round about the capital has been deindustrialized. Another possible explanation is that the existing advance factories are relatively large and, presumably, exclusive. Stockholm accounts for nearly half the aggregate property value (or rather, assessed value for taxation purposes) of advance factories (43%). The dominance (by value) of this region is also reflected by the median assessed values of different regions. For Greater Stockholm the median value for advance factories is barely SEK 9m for both Gothenburg and Malmö it is SEK 3.5m, and for the rest of Sweden it is not quite SEK 2m. This exclusive cachet is also reflected by many industrial premises being designed as offices and being applicable to both industrial and office use. But value is also conditioned by property investors paying more for premises in the big cities than premises outside them.

The ownership categories of the assessment units can repay closer study (Table 8.3). All the types of unit presented in Table 8.1 are also included in Table 8.3. Unfortunately it is not possible to present owner categories for different types of industrial property, nor the extent to which the different types are intended for direct activities or for letting. The test valuations preceding the 1988 property assessment, however, revealed that about one-quarter of all production facilities, industrial offices and warehousing facilities were partly or wholly tenanted.

Property ownership patterns of Swedish limited companies and physical persons differ as far as the number of units is concerned. Physical persons, as a rule, own small units, which makes the total value of their holdings more marginal. Joint stock companies are the big owner category in terms of value. The State, too, is a big owner.

Table 8.3 Owners of industrial units in 1988. Number of assessed units and aggregate assessed value by ownership categories.

	Number	Value (SEKm)
Physical person	23 161	11 355
State	971	68 087
Municipality	7 571	9 182
Church of Sweden	105	43
Estates of deceased persons	747	142
Swedish joint stock company	26 601	165 068
Housing co-operative etc.	1 241	3 376
Public utility	90	80
Other legal corporation	4 506	10 329
Ownership unknown	3 248	8 232
	68 241	275 895

(Source: SCB 1989b)*

* The same categories of industrial unit as in Table 8.1. The number of units differs between Tables 8.1 and 8.3, because these tables refer to different years.

8.3.2 Value development and production costs

The regularity of property assessments makes it possible to follow the development of regional real estate markets over time. The 1981 and 1988 assessments were based, respectively, on prices and rents in 1979 and 1986. During the first half of the 1980s there were a number of regions where industrial property capital was expanding. These included the metropolitan regions of Stockholm and Gothenburg and the regions around the university cities of Lund, Linköping, Uppsala and Umeå. Most county towns and large urban communities kept fairly well abreast of developments. Sparsely populated rural areas, and especially the interior of northern Sweden, by contrast, showed much poorer development for the period.

In 1979, the value of a given production facility was roughly 2.5 times higher in central parts of the Stockholm region than in the interior of northern Sweden. In 1986 it was ten times higher. It was above all the sparsely populated rural communities of northern Sweden which lagged behind, but differences between rural and central communities in each of Sweden's 23 counties have also changed adversely for the countryside. At the end of the 1970s, industrial units in the central localities of the counties were, on average, worth about 50% more than in the rural communities. Actual differences vary between 15% and 100%. Seven years later the corresponding intra-county differences had risen to something between 200% and 400%. Many industrial proprietors in rural areas of Sweden actually saw their property values stagnate during the period or else diminish in absolute figures (a situation compounded by 80% inflation). Owing to the poor development of property values in large parts of the interior of northern Sweden, a newly erected building has a technical present value equalling less than half its replacement cost (Sundquist, 1987; 1988). Developments since 1986 have done little to change this.

Turnover in the form of purchases of developed properties for manufacturing industry was about 1300 properties per annum at the beginning of the 1980s. Turnover then rose, exceeding 2000 in 1988. Roughly half these transactions were 'straight' purchases at market prices. The prices paid were relatively stable until 1984, after which they began to climb. To begin with, though, they fell in fixed money terms, but later this gave way to an increase. These developments are illustrated in Figure 8.1, which among other things includes an index for the sum total of purchase prices/assessed values of all 'straight' purchases per annum. The index is in fixed money terms (1985 = 10).

The reason for these price movements during the period may have been that demand for facilities fluctuated with the volume of industrial investments (Figure 8.1). In fixed money terms, the annual volume of investments in machinery and equipment declined during the first two or three years of the 1980s, after which it took an upward turn. Investments in buildings and industrial facilities, like the development of real estate prices, followed the same course but with a lag of one year or so. One should, however, beware of drawing unduly far-reaching conclusions from this figure, because property statistics are greatly influenced by the types of property sold. Nor has any allowance been made for regional factors.

Market turnover for building land for industrial purposes reveals a growth resembling that for developed industrial properties. The number of purchases rose from about 150 properties in 1980 to 400 in 1988. At the same time the average price of a development site rose from SEK 20/m^2 to SEK 90/m^2 (SCB, 1989c; and 1990b). There are great differences

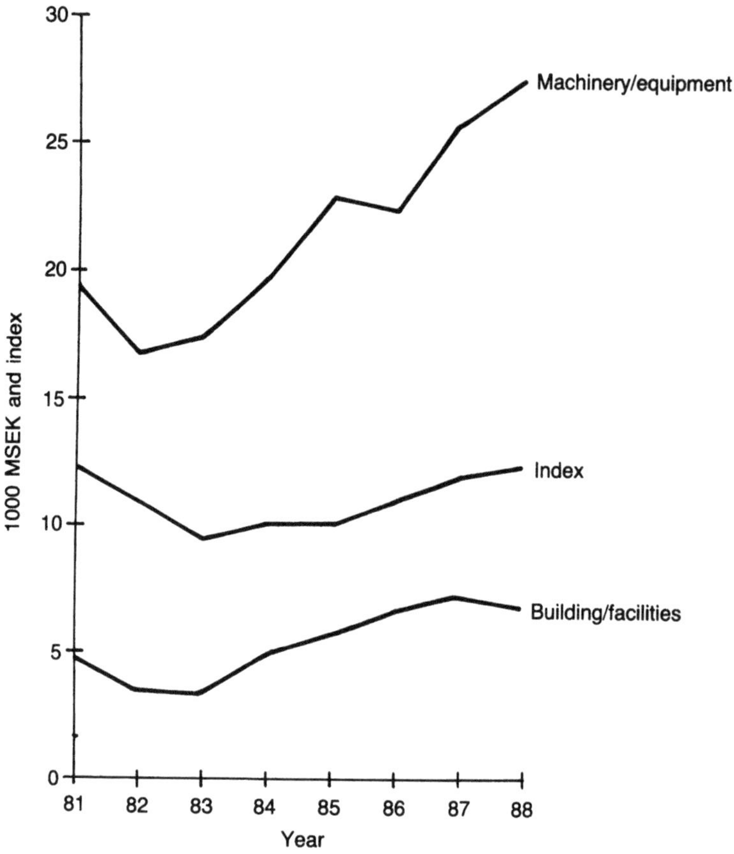

Figure 8.1 Total industrial investments in machinery/equipment and buildings/facilities, together with an index of purchases of industrial properties for manufacturing industry. 1981–89 in 1985 prices. (Source of calculations: SCB 1989c, 1990c and 1990d).

between regions. in 1990 the price of building land in the Greater Stockholm region seems to vary from SEK 100/m^2 to SEK 200/m^2 on the outskirts of the metropolitan region to about SEK 1000/m^2 in the inner suburbs. Land in medium-sized towns probably costs about SEK 50/m^2. In small communities the price is virtually negligible. Prices refer to land only, and thus do not include the cost of streets, water supply and sewerage.

Industrial building costs in 1990, planning costs included, are roughly SEK 7000–9000/m^2 for a simple industrial building with a steel carcase, sheet metal walls and insulation and a small office section. More office-like, better-designed industrial buildings can cost SEK 11 000–13 000/m^2, and for customized designs the cost of production depends entirely on the customer's wishes. To these figures must be added the cost of land, streets, water supply and sewerage and outdoor facilities.

The rents obtainable in the market barely correspond to these production costs, which frequently means a low return on capital employed, at all events in the short run. A property-management company has to be prepared for a deficit for the first five years after building, and must therefore have powerful capital behind it.

It is worth adding in this connection that the minimum rents charged by landlords for industrial and warehousing premises in the Stockholm region rose from SEK 400/m^2 p.a. in 1986 to SEK 580/m^2 in 1990. Actual rents vary, however, depending on the type of premises and their location. The average rent in 1991/2 is about SEK 700/m^2. Rents for office premises are two to three times higher (SML, 1991). In 1986 the rent of an ordinary production facility in the interior of northern Sweden was SEK 100–150/m^2 p.a. (Sundquist, 1988), and today it is perhaps in the region of SEK 300/m^2.

Annual rents can be estimated to include SEK 30–70/m^2 for running costs, while the remainder goes on capital yield. The direct return on the present market values of these properties varies from about 5% in good big city locations to 10% in poorer locations in the big cities and in the country at large. In the good locations, both market values and rents are higher, and so the lower return reflects assessments of the long-term security of the investments. By contrast, the direct return on current production costs appears to be roughly 5% in inferior locations and slightly more in better ones. The low return and deficient confidence in the future make it virtually impossible to erect industrial premises for rental purposes in, say, the interior of northern Sweden without regional support.

The return required on capital in industrial operations can make it an advantage to the industrial enterprise not to tie up capital in real estate, especially if rents are low. The variable rate of interest on bank loans to industrial enterprises (January, 1991) is roughly 13–14%. This applies both to property mortgages and chattel mortgages. Actual interest rates vary depending on customer relations and security. Floating rates of interest are geared to the general level of interest rates in the money market. The credit institutes advancing fixed-interest loans secured in real estate are currently charging similar rents for security in industrial properties (18%). Interest rates are often set for between one and five years. Up to 75% of the value of properties is mortgaged. Credit lines are lower for uncertain enterprises and in inferior geographic locations.

The growth of interest in sale and lease-back points to a tendency to segregate use from ownership, even though these transactions are primarily undertaken for financial reasons. Sale and lease-back has been in use for just over 10 years now, and the total market, i.e. not only industry, amounts to rather less than MSEK 40 000. Both industrial undertakings and local authorities are taking this opportunity to liberate dormant real estate capital without the risk of uncontrolled rent increases. The tenancy agreements are usually for between 15 and 25 years. The rent (which is really interest on the capital) relates to the general rate for day-to-day loans and at present is approximately 13%. It is often set with reference to financing, which means that the renting enterprise does not share in the appreciation or depreciation of the property. As a rule the property is maintained by the tenant, at his own expense. The first sale-back opportunity usually comes after 5–10 years, after which the option is normally an annual one.

8.3.3 Construction companies and property-management companies

For a long time now, the actual construction market has been increasingly dominated by the large construction companies. At present the five biggest companies control nearly two-thirds of the construction market. The big companies undertake both large and small projects all over the country, frequently spreading their production between housing, non-housing and heavy engineering projects. At the same time, many construction companies have been looking for new niches, due to the possibility of a dip in the construction market resulting from political decisions or other events. In this way they have become increasingly concerned with real estate management and finance, while still maintaining departments for building production. Management is closely allied to their traditional activities, added to which the existing stock of buildings may appreciate if a curb is put on building output, e.g. for political reasons. This way, corporate capital is shielded from rapid fluctuations in the outside world (Gorpe, 1989). The commercial property market has also attracted institutional investors, including pension funds and insurance companies.

Certain aspects of this income-producing real estate market are apparent from an interview survey by Brzeski *et al.* (1989) of 230 property companies. These included 14 capital-investing institutions (insurance companies and pension funds), 15 construction companies and 132 property companies. Seventy-five per cent of the companies looked on themselves mainly as property managers. Half of them had stocks valued at less than SEK 250m, and just over 20 had stocks exceeding SEK 2500m in value.

The survey shows that companies very often invest directly in real estate instead of putting their capital into mortgage loans. Thus they are banking on property management instead of capital placement. They distribute their property holdings between various types of property, such as offices, shops, housing and industrial premises. Just over 80% of the companies own office and retail properties, and almost the same proportion own housing properties. About 50% of the companies replying to the questionnaire have investments in industrial properties. There is a distinct tendency for small companies to specialize in industrial properties.

The tendency in investments is for office and retail properties to predominate, with housing occupying second place. Industrial properties come third. Investments in hotels/motels, agriculture and forestry and undeveloped land appear to be fairly uninteresting. Just over half of all companies have recently invested outside the metropolitan regions of Stockholm, Gothenburg and Malmö, mainly because the direct return on capital employed is higher outside the big cities. Other reasons are the lack of investment properties in the big cities and a desire to spread out property holdings geographically. Another striking fact is that 90% of the companies rely on their own knowledge of the market when making their investments. Only four out of ten carry out a direct market survey. Those doing so mainly investigate the development of property prices and, in the second instance, the local economic structure (Brzeski *et al.* 1989).

Thus, within the income-producing real estate market as well as the sale and leaseback market there is currently a tendency to separate legal title and use. Whether or not this will develop further in the future is probably dependent on factors such as property price, rates of interest, rent levels and rules of property finance.

8.3.4 Regional support

In conclusion, some mention will be made of regional industrial support (SFS 1990: 642). Such aid can be advanced to Swedish and foreign companies conducting or planning to conduct competitive business operations on commercial terms. This support, for example, can be provided for industrial operations and for the erection of small industrial premises (maximum 2000 m^2) for letting to several other companies. There are various types of grant, designed to facilitate start-ups, product development and job creation measures. Localization grants can be paid for the construction of new buildings and the enlargement and alteration of existing ones. These grants can cover up to 35% of total expenditure. The actual percentage will depend on location and project. Regional support is only available in clearly specified areas of Sweden within the sparsely populated regions of northern Sweden and the forest communities of central Sweden. As a temporary measure, however, the Government can expand the regional support areas to include areas affected by extensive structural changes. Applications for regional support are processed by the county administrative boards and the National Industrial Board (*Statens Industriverk*).

8.4 CASE STUDIES

8.4.1 Kummelberget

Kummelberget is a 25 ha industrial estate for light industry and crafts, interspersed with warehousing and offices. It is located 12 km east of central Stockholm and 2 km from the centre of Orminge. The estate was developed on a greenfield site previously belonging to the Municipality of Nacka. The industrial estate has 75 lots, most of them developed. These are either freehold or site leasehold. Some buildings have rental facilities. There are also instances of sale and lease-back. Several buildings are advance factories. Demand for properties and facilities is high, and so too is turnover.

The first municipal detailed plan for Kummelberget was adopted in 1966, and since then it has been successively enlarged and revised. The latest major alteration, in 1985, was intended to create opportunities for new start-ups. A draft plan was prepared but was modified after consultations, having been criticized above all for insufficient provision for nature conservation in the surroundings. The new detailed plan was adopted in 1987. The very next year it was expanded eastwards, to make possible the development of another industrial estate and to improve the economic basis of the development plans.

The planning regulations prohibited the establishment of environmentally disruptive industry. Furthermore, certain properties had to be surrounded by fencing at least 2 m high, so as not to disfigure the surroundings. Building heights were regulated for various sub-areas, the maximum permissible height being 11 m. The plan also laid down that the construction and maintenance of streets and of water and sewerage mains was a municipal responsibility. Connection to district heating, on the other hand, was to be optional.

The planning regulations did not include any stipulations on the size of plots. The land being municipally owned, plot sizes were regulated in the process of the municipality selling off its land or leasing it at a ground rent to various entrepreneurs. The principle applied to freehold conveyance was that, in the central and hitherto untouched parts of the area, plots of about 3000 m^2 were to be created for companies in need of presentable facilities. Elsewhere in the area, more space-demanding enterprises were to be located on plots varying between 5000 and 15 000 m^2. One-third of the area was reserved for plots of less than 1000 m^2.

The entrepreneurs have a free choice between freehold and site leasehold. Both freehold prices and site leasehold ground rents are set at market rates. The site leaseholds run for 40 years, with 20-year extensions. Ground rents are revised every 10 years. The current ground rent is SEK 30/m^2 p.a. for land, up to 50% of which may be built on. The current freehold price of land is SEK 1000/m^2 building rights. Prices include street connection charges, but they do not include water and sewerage connection. The charge for the latter is roughly SEK 100/m^2 floor space. District heating, which is privately operated, is subject to an additional connection charge if required. The electricity connection charge is negligible in this context. We may add that annual rents per square metre are about SEK 600 for industrial facilities and about SEK 1000 for offices.

The area includes a number of small, modern industrial facilities. One of them is Bo Instrument's offices, workshop and warehouse (Figure 8.2). This land was sold off by the municipality in 1986 and parcelled in connection with the sale. Facilities totalling 850 m^2 were built on it immediately afterwards. The building permit procedure took about three months and the construction time was six months. Another 350 m^2 of high-bay warehousing were added in 1990. At present the building is being used for the company's own production and sales, which include instruments for gas analysis and detection.

Not far away there is a craft estate comprising 14 different properties in four enfilades (Figure 8.3). Building rights here emanated from the 1987 enlargement of the detailed plan, following which the Municipality of Nacka sold the area to a development enterprise, with the stipulation that building development was to comply with a given sketch of terrace

Figure 8.2 Bo Instrument's building, combining office, manufacturing and warehousing facilities.

Figure 8.3 The crafts village on the Kummelberget industrial estate. The enfilade contains four separate properties for light industry. The plots, like the building areas, are slightly less than 300m² per property.

factories for light industry and offices. If the area had not been developed by 1990 at the latest, a contingent fine would become payable to the municipality. The development company engaged a firm of architects to plan the buildings and a special construction company to co-ordinate the building development. The construction company purchased and conducted the development by engaging subcontractors for the various building operations. Building took place in 1988/89, after which the 14 properties were parcelled off and sold by the owner. Some purchasers acquired more than one property each. Selling prices were for the most part between SEK 2.5m and SEK 3m per property. Plot sizes varied between 250 and 350 m^2 and plots were almost entirely built over. Upwards of 5000 m^2 was reserved for communal parking and communication spaces. The newly formed properties are used both directly and on a rental basis.

Another crafts village for sale and rental occupation is under construction at Kummelberget, using a development technique very similar to that which has been described.

8.4.2 Edsbacka

So far we have been dealing mainly with strictly industrial properties. The fact is, however, that quite a lot of modern light industry is located in office-like complexes, often right next door to offices. The technique of exploiting this type of property is very similar to the development of industrial properties, as can be seen from the example of Edsbacka in the Municipality of Sollentuna. Edsbacka is a small, well-designed enterprise zone covering 60 000 m^2. It is located 15 km north-west of downtown Stockholm and about 2 km from the local municipal centre. The area used to be a gravel pit.

The municipality, which had owned land here for many years, began drawing up a detailed plan at the beginning of the 1980s, and a plan was finally adopted in 1986. About 50 000 m^2 were designated for building development, the remainder being reserved for municipal streets and for water mains and sewerage. The planning provisions earmarked the development sites for offices, factories and warehousing. A petrol station and motel were also sanctioned, within a specially designated part of the plan. The plan stipulated that building rights were not to exceed the total development acreage. Building heights were restricted to 10, 14 and 20 m respectively for various sub-areas, but there were no stipulations concerning the more detailed design of buildings. Instead, matters of this kind were to be decided through the award of building permits.

Subsequently the municipality conveyed sub-areas to various interests, including the sale to a private development enterprise (Industriprojektering AB) of a plot measuring about 11 000 m^2. The contract of sale stipulated that the buyer was to build office and service facilities, divided into several buildings. These were to be of brick and were to have a combined built area of at least 2000 m^2. If this minimum area had not been built on before 1989, a contingent fine was to be imposed. The purchase price was market-adjusted and inclusive of street, water and sewerage connection charges.

Instead of acquiring the freehold immediately, the developer had to make do with a beneficial interest agreement. The land was not conveyed in freehold until the bottom slabs of the buildings had been laid down. This conveyance technique prevented the developer from mortgaging the property during the introductory phase, forcing him to use equity instead. From the municipality's point of view, the arrangement provided an additional guarantee of the area being developed. Twelve properties were formed and one building erected on each of them. The land in between the building volumes was made common to the properties, for use as communication spaces, parking facilities and parkland. Communal facilities were also established for lifts and stairwells and for the garaging constructed beneath the large, continuous building volumes.

The developer constructed the buildings in two stages between 1986 and 1989 (Figure 8.4). About 8000 m^2 of rental space was constructed (gross area 12 700 m^2) in buildings of two or three storeys. Building permits were obtained within just over a month and each phase took rather more than one year to complete. Building was financed partly by loans on the security of the properties and partly out of equity. Building operations were conducted on an in-house basis, with the developer purchasing construction services from different contractors. Seven properties were retained within the developer's conglomerate (the associated company IP Contracting AB) to be let to other companies, while five were sold off individually to various interests. Two of these buildings are not let, while three are owner-occupied.

The buildings are designed for optional use as offices, manufacturing/service facilities and warehousing. Offices currently occupy about 70%, and this figure is tending to rise. Rents are in the region of SEK 1300–1400/m^2 per annum.

Figure 8.4 Edsbacka. These facilities comprise 70% offices, the remainder being used for services, crafts and warehousing.

REFERENCES AND FURTHER READING

AB72, ABT74, AMA, Mark AMA etc. (standardized agreements and regulations for entrepreneurs). Distribution Svensk Byggtjänst, 17188 Stockholm.

BFS 1988: 18(1988) *Regulations and General Recommendations for New Building Constructions*, Allmänna förlaget.

Brzeski, W. J., Jaffe, A. J. and Lundsbrom, S. (1989) *Commercial Real Estate Investment in Sweden*. Dept of Real Estate Economics, KTH, 100 44 Stockholm.

Gorpe, P. (1989) *More dwellings, less premises*. Länsstyrelsen i Stockholms län, rapport 1989: 10.

SCB (1989a) *Statistisk årsbok 1990*, Statistics Sweden, 115 81 Stockholm.

SCB (1989b) Statistical Data Bo38SM8901.

SCB (1989c) Statistical Data P18SM8901.

SCB (1990a) Statistical Data Bo37SM9001.

SCB (1990b) Statistical Data P16SM9002.

SCB (1990c) Statistical Data F13SM9002.

SCB (1990d) Statistical Data P15SM9001.

SFS numbers refer to the Swedish Code of Statutes (Svensk författningssamling). All SFS references are to enactments printed in *Sveriges Rikes Lag*, the statute book published annually. Ordinance SFS 1990: 642 on Regional Support to Enterprise, however, is not included.

SML (1990) *The Market of Premises in Stockholm* Stockholms Mark- och Lokaliseringsbolag, Hantverkargatan 5, 112 21 Stockholm..

Sundqvist, A. (1987) New Property Assessment. *Svensk Lantmäteritidskrift*, No. 3.

Sundqvist, A. (1988). AFT 88 – No. taxering av industrienheter. *Svensk Lantmäteritidskrift*, No. 2.

Sundqvist, A. (1988) *AFT 88 – Assessment of Industrial Property*. Svensk Lantmäteritidskrift, **2**.

9

THE UNITED KINGDOM

Barry Wood and Richard Williams

Much has been written elsewhere about the UK planning system, and there are also several texts available on the operations of the property and real estate markets. Consequently this chapter seeks to offer introductory guidance and explanation of key principles and distinctive features of the industrial property market in the UK, and to convey a sense of the style of operation and forms and extent of government intervention, so that a non-British reader is better equipped to use texts written for a British readership and interpret the greater detail to be found there.

This chapter is structured as follows: the legal framework is outlined first, concentrating on landlord and tenant and property Acts, and the Town and Country Planning Acts. Secondly, actors and agencies in the planning and property market, in the public, private and professional sectors, are reviewed, as this is an important basis for understanding how industrial property development operates in practice.

Having established the legal framework and the actors in the process, we go on to discuss the market in industrial property, associated public sector policies, programmes and planning procedures, and present two illustrative case studies. These both have an urban context, but from very different market conditions: Stockley Park near London's Heathrow Airport and Team Valley Trading Estate in Tyneside. The chapter concludes with a synthesis and suggestions for further reading.

A further preliminary is also necessary before proceeding to the main part of the chapter. The chapter is intended to describe the whole of the UK. However, although the UK has a unitary rather than federal constitution, it has four legally distinct parts: England, Wales, Scotland, Northern Ireland. The general principles and institutions of the industrial property market apply throughout the UK, but the details of law,[1] procedure, regulation and subsidy vary. In general, English law is described in this chapter. This also applies to Wales although it is administered by a different Government department. The Scottish legal system is quite separate, and different Acts of Parliament apply to Scotland, although their purpose is usually the same as the English equivalent. Northern Ireland is under direct rule, and local government there plays no part in planning as it does elsewhere in the UK, the Northern Ireland Office being the executive authority. The basic principles remain as in England, as is also the case with landlord and tenant law.

9.1 LEGAL FRAMEWORK

There are three sets of legislation providing the framework within which the industrial property market operates: landlord and tenant and law of property Acts; the Town and Country Planning Acts; and the Finance Acts. Key concepts and operating principles of the first two of these are outlined here, but the Finance Acts are beyond the scope of this text.

The legal framework surrounding the occupation of industrial property can be usefully divided into that associated with occupation via property ownership and that associated with tenancy. An owner-occupier is one who buys the freehold (or outright ownership) of a property which is not occupied by a tenant with a legal right to maintain that occupancy. The owner can then occupy and make use of the premises on the basis of its existing use. The freehold method of tenure generates the most complete collection of property rights available in the British system. In addition to occupation the owner can let the property to tenants, sell it, use it as security for a loan and generally even allow it to decay. The rights are not absolute, however. Planning legislation limits the rights of change of use, demolition and redevelopment.

In principle, the owner-occupier who buys the freehold of an industrial property with the intention of using the property for an industrial activity, and in the absence of an existing tenant with occupation rights, will be able to occupy the property indefinitely with no further charge other than the Uniform Business Rate (the UK's non-residential property tax). The freeholder can also apply for planning permission to redevelop the site and will gain any increases in land values associated with such permission. The owner has the right to sell the property, but the rights to change use within the building are restricted to the use class which the building has been given, and planning permission for a change of use is required for other uses.

Freehold ownership is attractive to industrialists because in addition to the above benefits it is also possible for a new property to be built on the site in accordance with the specific requirements of the firm (subject to planning permission). Occupation by means of a tenancy is also popular in Britain. The legal framework controlling business tenancies is mainly derived from the Landlord and Tenant Act 1954 and the Law of Property Act 1969. These Acts apply 'to any tenancy where the property comprised in the tenancy is or includes premises which are occupied by the tenant and are so occupied for the purposes of a business carried on by him or for those and other purposes' (Law of Property

Act 1969). While exceptions to this wide-ranging definition exist, most notably agricultural tenancies which are controlled under different legislation, the Acts incorporate almost all industrial activity. A tenancy is created in the form of a lease which is a legally binding contract between the freeholder and the occupier. The tenant can remain in occupancy for the finite length of the lease as long as the terms of the lease are fulfilled. The tenant's rights are in practice restricted to occupation.

Under the Town and Country Planning Acts, all land uses are classified under the Use Classes Order. A change of use class requires planning permission whereas a change of use within the same use class is permitted. In practice the uses which the owner permits within the lease are likely to be more restrictive. For short- or medium-term leases the tenant is usually required to pay for the insurance and maintenance of the building for the term of the lease, and to pay the agreed rent (quarterly), which is determined by market forces in the absence of government control. With a 21-year lease a fixed rent would suffer at the hands of inflation, and consequently modern leases include a clause allowing for an increase in rent levels, up to the current free-market levels at regular intervals. Seven-yearly reviews are common and five-yearly intervals are also often available. Where owners and occupiers disagree on the appropriate rent increase then there is a well-established procedure by which an independent expert determines the new rent.

In addition to the rent the tenant is required to pay a service charge for those facilities provided by the landlord. For example, where buildings are occupied by more than one tenant, then much of the common space, corridors, entrance hall, car parks etc. will remain the responsibility of the landlord but the costs of operation and management of the facilities will be passed fully to the tenants via service charges. If the tenant breaks the terms of the lease then the freeholder can apply to the courts for recompense, and ultimately the freeholder can force a defaulting firm into bankruptcy.

The freeholder (or landlord) can sell their interest in the property during the term of the lease, and indeed the tenant can assign the lease (sell the tenancy rights) to another tenant if that is feasible and desirable (from the tenant's viewpoint). This means that the lease structure is not as binding on owners and occupiers as it at first appears. For landlords in particular the ability to sell the tenanted property implies that a property investment can be converted back to cash more easily (though there is still the possibility of conversion at a capital loss). In practice this process has substantially facilitated the development of the property investment market. At the end of the lease a freeholder may offer a new lease or give the tenant a notice to quit. If the tenant does not wish to quit then the freeholder will need to show a court that there are clear grounds for opposing a new tenancy. Such grounds are well defined in law and include the freeholder's need to obtain a vacant site for property redevelopment purposes. The lease creates a legally binding obligation on the tenant to ensure that the owner's property rights are fully protected. In practice these obligations are not onerous but their existence does emphasize that prospective tenants should obtain legal advice.

The landlord-and-tenant arrangement is popular with both property investors and tenants. For investors it offers the prospect of a secure, and not completely illiquid investment offering a stable income flow with regular updating to take account of market pressures; and it is to some extents inflation-proof. For tenants it offers the opportunity to obtain premises without the outlay associated with the purchase of the property. It is seen as particularly valuable for new and rapidly growing small firms for whom property ownership would be very burdensome.

An additional form of tenancy occurs where a long lease is created in exchange for one initial payment (known as a premium). At the end of the period the property reverts to the original owner. This form of tenure is a mixture of landlord/tenant and owner-occupier. It offers advantages to the landowner in that a large return can be immediately obtained while the benefits of eventual redevelopment are preserved. The owners can even sell their long-term interest in the property. For the occupier the risk of rising rents is avoided and occupation of the premises can be assured. In effect, though not in law, the tenant becomes an owner-occupier for a short period of time. The tenant can sell the occupancy rights to another business, and this facility has been widely used, with the result that there is now a substantive market in shorter leases for which an initial payment must be made.

The process of selling tenancy rights is known as an assignment (or a sublease where the agreement is for a period shorter than the remaining life of the lease). Assignment rights may be restricted by the lease, and do not eliminate the first tenant's ultimate obligation to fulfil the terms of the lease if the second tenant were to default. Legal advice should be sought about this issue. The long-term lease in exchange for a premium payment is particularly popular with well-established companies who expect to be long-term occupiers of the site.

The occupation of property entails compliance with a range of other legislation. The most significant Acts are:

1. The Health and Safety at Work Act 1974;

2. The Employment Protection Act 1978;
3. The Environmental Protection Act 1991.

Again the importance of legal advice must be emphasized. The legal framework surrounding property investment is outside the scope of this book. Professional advice is available from the major surveying firms, lawyers and (increasingly) financial management consultants.

Town and country planning legislation in England and Wales has been made somewhat simpler recently by the passage of the Town and Country Planning Act 1990 (TCPAct), which is a consolidating Act incorporating in one statute most of the legislation which currently applies (and all that relates to questions of land use allocation and authorization of development for industrial property). The object of such legislation is to ensure that all the basic planning law framework is to be found in one place. Already, however, further legislation in the form of the Planning and Compensation Act 1991 has been passed.

Industrial development projects may also take advantage of other powers such as those in the Inner Urban Areas Act 1978, which established the basis of inner city policy; or the Local Government Planning and Land Act 1980, which contains the legal powers on which enterprise zones and urban development corporations are based. In this section, certain basic principles of the TCPAct only are described, but use of these other Acts is illustrated in the Team Valley case study below.

The system of control of development has followed the same principles, although with many changes of detail, since the 1947 TCPAct which laid the foundation of the post-war planning system. The basic principle is that development is given an all-inclusive definition, and all operations that constitute development require planning permission unless an explicit exemption applies. It is worth quoting in full the definition of development inherited from the 1947 Act: 'the carrying out of building, engineering, mining or other operations in, on, over or under land, or the making of any material change in the use of buildings or other land' (s 55, TCPAct 1990). This is of course subject to supplementary definitions and covers most imaginable possibilities.

Applications for planning permission are submitted to the local planning authority (LPA). Normally this is the district or municipal council for the area in which the proposed development is located, although it could be a development corporation (see below) or national park authority. The LPA has eight weeks in which to determine an application (16 weeks if an environmental assessment is required under EC Directive EEC/85/337), and may grant permission with or without conditions, or refuse permission. Central government intervenes only in a minority of cases, either if a proposal is a major departure from existing policies or if the applicant appeals against refusal, conditions or failure to issue a decision in time. When an appeal is made, a planning inspector is appointed who either holds a hearing in public, known as a public inquiry, or investigates the case by means of written submissions. A decision is made based on these findings by the inspector or in a few major cases by the Secretary of State.

Consideration of a planning application is an important power, and LPAs have discretion to decide applications on their individual merits. This offers valuable flexibility in the planning process, but has been criticized for loss of certainty on the part of the applicant or developer. The LPA does not operate in isolation, however. It must take into consideration any development plan for the area, the views of other organizations and statutory authorities who have to be consulted, and the public acceptability of its decisions.

This short introduction to the planning framework has started with the development control system because this aspect has the highest profile for any prospective industrial developer. Development control operates within the context of a system of development plans. All counties (regions in Scotland) must have a structure plan, approved by the appropriate Secretary of State, setting out the strategic policies for the economic and physical development of the county (region), and reports of survey and public consultation on which these policies are based. Structure plan policies are expressed in verbal and diagram form, but explicitly cannot be expressed in precise land-use terms on a cartographic map because they are strategic documents.

Most urbanized parts of the country also have adopted local plans. Following the passage of the latest legislation, the Planning and Compensation Act 1991, local plans are to become mandatory in England and Wales as they already are in Scotland. Local plans indicate in site-specific land-use terms the manner in which the LPA wishes to see land developed, and are therefore a guide to applicants and developers. Outside Scotland, however, some areas of potential industrial development may still not have a local plan.

In deciding planning applications the LPA must take into account the structure plan and any local plan, and be able to justify any departure from their land-use allocations. The LPA is not bound to follow them, however. It may refuse permission for development in accordance with the plan, or approve proposals not envisaged by the plan. This is the essential feature of the UK's discretionary system, distinguishing the UK planning system from that of most other European countries.

Such a discretionary system could not function efficiently without policy guidance because it is necessary both in producing the plan and in guiding development control decisions. Part of the guidance is provided by the Secretary of State at the point where the plan is produced. Advice is given on the form of plan before it is produced and ultimately strategic advice has to be incorporated within local plans. Structure plans have to be approved by the Secretary of State for the Environment and this enables central government to maintain both control and co-ordination.

The Secretary of State can also effectively change guidance by the production of circulars which modify ministerial advice. For example, the Secretary of State has recently informed the local planning authorities that when examining open-cast coal-mining applications they should no longer consider whether there is a need for the coal but rather they should base their decisions on the environmental effects that such mining would have. This policy guidance overrules whatever is written in the local plan. This policy-making process has generally been seen as a top-down, central-government-led activity. Not all policy has been formulated in this way, and historically the Department of the Environment has been prepared to reformulate public policy in the light of local government and pressure group representations.

Generally the practice of plan-making has ensured that, where possible, sufficient land is made available for the industrial property sector, though this may not be specifically for industrial users. Problems have only arisen in the most buoyant areas where demand for land by other users has resulted in inevitable conflicts for plan-makers.

Most other industrial property policies have evolved as part of regional policy. Ever since the high-unemployment years of the 1930s successive governments have subsidized industrial property development to attract new firms into the 'poorer regions' as a method of overcoming regional disadvantage. The Team Valley case study examines one of the methods in more detail. The essential point is that government has not been prepared to let what it sees as the slow process of market adjustment determine the availability and price of industrial property.

In the 1980s government policy towards the regions has changed, with an emphasis on encouraging efficiency and wealth creation. Some public agencies in the field have been forced to reduce their activity (e.g. local authorities) while others have been privatized (e.g. new town development corporations). The most important agency, English

Estates, remains an active participant in the development process but the use of enterprise zones and urban development corporations (see below) points to both a change in policy to a market-led approach and a political statement by the current government. Another pro-market policy change, and one which covers all of the country, has been the change in the Use Classes Order which effectively permits land with planning permission for industrial property usage to be utilized for general business use, including a substantial element of office space. This policy change has had most impact in the more prosperous regions of the country where the demand for such space is able to outbid traditional industrial users.

The planning system is sufficiently flexible to allow for the possibility of developers' offering benefits to the local community in exchange for planning permission in areas where normally planning permission would not be granted. This is generally known as planning gain, although since the passage of the Planning and Compensation Act 1991 the new term 'planning obligation' is to be used. Planning gain has often been a contentious issue, and the new law concerning planning obligations is more detailed and precise. Generally, developers of industrial space have not been in a position to offer planning-gain benefits to the local community, for industrial property development has simply not been profitable enough to finance such activity. In the 1980s, however, the new Use Classes Order has created a new class of up-market office parks where the development gains would be sufficient to finance such social benefits. The case study of Stockley Park is an example of such a development though its planning permission was actually achieved before the creation of the new Use Classes Order.

Overall, the policy framework has generally been one of assuring that sufficient land has been made available for industrial development within the plans, and the element of discretion available to local authorities has not been as important here as in other property markets. Coupled with public agencies providing subsidized space, this policy has assured that industrial property development has taken place in accordance with the plan. In the 1980s policy has become more market-led and industrial property development in the regions has been heavily influenced by the subsidies within enterprise zones. Within the more prosperous areas the planning system has been forced to become more responsive to the locational requirements of industrial property developers.

9.2 ACTORS AND AGENCIES

9.2.1 Regulatory

Local authorities play a key role in the authorization of development, as they frequently do in the process of assembly of land for industrial development and the promotion of urban development schemes. Therefore local authorities are inescapably involved as key actors in the development process, so it is important to keep in mind an outline of the structure of government in the UK.

The power to legislate for the whole of the UK, or for any part of it, is held only by the national Parliament in London. Executive powers are held by central and local government. For industrial development, the major government departments responsible for overseeing the system and proposing legislation are the Department of the Environment and the Department of Trade and Industry. These have jurisdiction in England only. Equivalent powers are exercised by the Welsh Office in Cardiff and the Scottish Office in Edinburgh. The Northern Ireland Office in Belfast does not merely have a supervisory role: it is also the executive authority responsible for planning policy and planning permissions there, as local authorities have no planing powers in Northern Ireland.

In the rest of the UK, local authorities have an important executive role in land-use planning, economic development and the consideration of planning applications. London and six other major English conurbations have a single tier of local government: London boroughs and metropolitan boroughs. Orkney, Shetland and the Western Isles of Scotland also have single-tier island authorities. Elsewhere, there is a two-tier system of local government: counties and districts in England and Wales; regions and districts in mainland Scotland. This structure is once again being reviewed by the government.

All of these authorities are local planning authorities (LPAs) except the districts in the three most rural Scottish regions. In general, planning applications are decided by the districts although the advice of the county/region is sought, and may be decisive in the case of major industrial development proposals. The UK has no regional government or authorities at the UK regional level (EC levels, I, II): Scottish regions should be regarded, in European comparative terms, as local authorities.

Local authorities consist of councils elected on a four-year cycle. The council has certain powers and duties assigned by national legislation, and exercises these powers through committees (e.g. planning, economic development) advised by permanent officials. Planning officers are responsible for examining planning applications, ensuring procedures and consultations are properly undertaken, and advising the committee on the planning merits of the proposal. The committee then makes the decision.

It is possible for a body other than a local authority to exercise planning powers. The main circumstance in which this arises in the context of industrial development is where planning powers are exercised by development corporations. These have boards appointed directly by the Secretary of State, not elected councils, and are established where a major concentration of planning and development effort is required. New town development corporations have been responsible for the new town programme, and more recently urban development corporations (UDCs) have been set up in certain inner city areas to promote new development, exercising all the TCPAct powers within their designated area.

Finally the regulatory power can be pre-set by central government by the designation of a zone where development is deemed to have permission as long as the form of the proposed development is in accordance with the rules for the zone. In both simplified planning zones and enterprise zones the regulatory role of the planning authority is to check that all development proposals are in accordance with the zone's rules and only to determine a planning application for developments which are proposed in violation of the zone's rules. Enterprise zones are created for 10 years during which both property development and occupation receive subsidies. Simplified planning zones also exist for only 10 years but differ in that there are no subsidies available.

It is actually possible for planning authorities to declare zones where such a permissive regulation would exist but such a declaration would be a policy statement and would not have statutory backing. Planning applications would have to go through the same process as all other applications but there would normally be more certainty over what would be approved on the particular site. These are the exceptional circumstances where the British system functions in a similar way to most continental systems with their statutory plans.

9.2.2 Professional advisers

The major professions responsible for the operation of the industrial property market are the real estate and planning professions. Accountants, lawyers, architects and financiers are also important. Professions in the UK are incorporated in non-government institutes, usually licensed by Royal Charter to regulate qualifications for entry, professional practice and ethics, for their profession.

The real estate profession is incorporated by The Royal Institution of Chartered Surveyors, which has approximately 51 000 members. They are entitled to use the professional title 'Chartered Surveyor' and the designatory letters ARICS/FRICS (Associate/Fellow of the RICS). They offer a wide range of surveying, building, property development and valuation expertise. This wide range associated with one qualification is unusual in real estate in other European countries. Members may offer professional advice as independent liberal professionals (fee-paid), or may be employed by property development companies or local authorities (salaried). The majority are in private employment.

The planning profession is similarly incorporated by the Royal Town Planning Institute, which has approximately 15 000 members, and may use the professional title 'Chartered Town Planner' and the designatory letters MRTPI/FRTPI (Member/Fellow of the RTPI). Most are in LPA employment, although many work as independent fee-paid professionals, often in association with surveying practices, with developers as clients.

In the UK context, there is no contradiction implied by being at the same time a professional adviser and member of a chartered profession, and a salaried employee of an LPA or other public body. The duty remains the same: to offer the best professional advice to their employers, whether they are private clients or local councillors.

9.2.3 Operators/entrepreneurs

The operation of the real property market is determined by the behaviour of certain key actors: landowners, developers, financiers, investors and property users.

Landowners

Land ownership in Britain is usually discussed in terms of freehold ownership where the rights in land are held in perpetuity. Before the industrial revolution of the late 18th and 19th century land was commonly held in very large units by a very small number of people. Some traces of this ownership pattern exist, particularly in rural areas, but much urban land has now been transferred to the ownership of industrial and financial businesses and government. The private owners are predominantly interested in obtaining the best returns possible from their land, and the land is considered as an investment. Profitable development is the normal investment route, but in the absence of investment opportunities, the land may remain unused. Additionally, land may be held unused for considerable periods in anticipation of future investment opportunities perhaps brought about by a change in the regulatory regime or perhaps by a new willingness of planning authorities to use their discretionary powers.

Public ownership of land is extensive in many urban areas. Much of it was obtained by local authorities in the 1950s and 1960s by compulsory purchase orders which were used to obtain sites for housing redevelopment. Other land was obtained when local authorities acted as partners in town centre redevelopment schemes. Other public bodies also own land, most notably British Rail and the Ministry of Defence. Another major public landowner, recently privatized, is the water industry. During the 1980s the public authorities curtailed their purchases of land and have been progressively encouraged to sell surplus holdings. While public bodies have been widely criticized for underutilizing their land holdings, much has changed in the 1980s. British Rail, for example, has started a process of redevelopment of many of its central London stations, each of which involves the use of some of the land for office building. This commercial approach is now common, though some public bodies, most notably local authorities, may introduce non-commercial factors into their land dealings.

Property developers

Property developers are those private or public organizations which coordinate the processes by which a new property is created. As this usually involves risk, these organizations are invariably (though not inevitably) entrepreneurial. Property developers may be landowners, builders, investors or even eventual occupiers but none of these attributes is essential. Private developers vary in size from firms with a handful of employees to major multinational organizations with billions of pounds (£) of assets. Public agencies include local authorities, new town development corporations, urban development corporations, English Estates and the Scottish and Welsh Development Agencies.

The private-sector property companies are the operators who seek out development opportunities, in terms of demand levels, which will ensure that development will be profitable. They have to obtain land, planning permission, finance for the project and supervise the construction of the premises. Upon completion they will be responsible for letting to tenants and/or selling the property to an investor. In practice the development arrangements can be immensely complex. For example, a developer may go into partnership with a local landowner. Planning permission may involve a legal agreement with the local planning authority to provide social benefits for the community. Finance for the development may come from a bank in the form of a loan with flexible interest rates, and the property may be built speculatively, i.e. in anticipation of obtaining occupiers when the premises are complete. In practice successful property development is highly profitable but it also involves many risks. In times of economic depression and high interest rates property development is one of the sectors most at risk of corporate failure.

Public-sector developers also take risks. They face essentially the same problems though they usually only attempt to develop land where planning permission already exists or is highly likely to be given. Public actors also have the resources with which to withstand poor returns or complete failure. In a previous section the importance of these agencies as a part of regional policy was noted.

Financiers

Financiers are those organizations which provide funding for the process of property development. Borrowing via the Stock Exchange (equity finance) has been a major source of funds for property companies. Finance has been attracted from the complete range of individuals and businesses who invest in new share issues, and as shareholders, they both take the risk that the property development may fail and earn any profits if it is successful.

Borrowing via fixed-interest loans (debt finance), is available from the major banks and many specialized financial intermediaries. This form of borrowing is popular with property companies, for it permits the profits of property development to be shared by a smaller number of shareholders. Lenders take the risk that the property development will be successful but the risk is moderated by limiting the loan to at most 80% of the projected cost and by requiring some form of security for the loan. Generally, banking principles are upheld: if the borrower wishes the bank to lend at greater risk then the interest rate charged will be higher and the banker will seek greater asset backing for the loan.

Finance for property development in the past has been provided by major insurance companies and pension funds but these institutions have never been the dominant actors in the industrial property market and in the later 1980s they have been almost completely absent from the sector. During the 1980s (up to spring 1989), lending for property development was highly competitive. Since then the downturn in the UK economy and the associated rise in interest rates has caused widespread depression in the property-development sector and new finance is both expensive and difficult to obtain. Finance for property development in the public sector comes from the government or via borrowing from the private money markets. In the latter case the loan would be guaranteed by the government. The various public agencies face a variety of controls and conditions on their borrowing and these are subject to change over time.

Investors

Investors in property are those who purchase property for the returns which can be obtained from it. While the range of property investors is immense, and includes the Church, private individuals and industrial companies, the major actors are the pension funds, insurance companies, property companies and government agencies.

Pension funds and insurance companies are commonly, though somewhat misleadingly, called the financial institutions. They both obtain long-term deposits from their clients and seek profitable outlets for them. Property generally has been one of the major investment sectors, (with equities and government securities), though industrial property has not been popular with them since the late 1970s. This is in part because of the poor performance of Britain's industrial sector and also because of the greater management costs associated with such property.

Property companies may invest in property as well as create it via development. They seek financial returns from property but, unlike the financial institutions, they are prepared actively to manage their investments and generally incur more risk. Property companies are major investors in industrial property and are active buyers and sellers of these assets. They obtain advice from the surveying profession but also employ surveyors directly within their organizations.

The government agencies which undertake development also hold property as investors. During the 1980s they have been increasingly required to obtain the maximum return from their investments. This is in contrast to the 1960s and 1970s where every effort was directed towards encouraging employment in the depressed industrial areas where these agencies are active. In recent years the new town development corporations have been progressively abolished and their industrial property assets sold to private investors.

In discussing the operators/entrepreneurs who operate in the industrial property market the central role of the commercial property companies should now be apparent. They are not only the prime co-ordinators of the property-development process but are also major buyers and sellers of the existing property stock. They are the key risk-takers in what is a high-risk market. When the property market goes through an upswing of an economic cycle they are normally very profitable businesses but at other times the smaller firms often fall into bankruptcy. While the surveying profession is widely held in high public esteem the property developers are commonly perceived as somewhat discredited. In part this unreasonable view reflects a lack of public knowledge of the activities of property companies; in part it reflects on the dubious behaviour of a few firms; but mostly it reflects on the British attitude to successful businesses.

Users

The users of industrial property, whether owners or tenants, vary from one-man businesses to the industrial activity of major multinational companies. In terms of the British industrial classification of business the occupiers come mainly from the manufacturing and distribution sectors.

Each business needs an enclosed space with minimal facilities other than heating and lighting, an electricity supply, a small office space and other facilities as specified by the Health and Safety at Work Act and other legislation. These minimalist requirements by occupiers ensure that each property can be used by a variety of businesses and this facilitates the existence of a real British market in industrial property. During the 1960s and 1970s the investors, particularly the financial institutions, insisted that developments were created to a very standardized design so as to ensure that they could be re-let at a later date to as wide a range of businesses as possible. These design criteria tended to ensure that the properties were good investments but they were not necessarily ideal properties for users.

Changes in user demand in the 1980s and the response of the market are considered below.

The general public

In a democratic society, the development process, for industry or any other sector, must be carried out in a manner acceptable to society as a whole. It is for this reason that development authorization is in the hands of elected authorities. There are, however, two other important ways in which public acceptability is tested. One is the process known as public participation, required by the TCPActs when new plans are created, and the other is through the operations of pressure groups. These are any association or organization which seeks to influence decision-making without actually assuming the powers of government itself. It is important to note that in Britain's discretionary system pressure groups attempt to influence decision-making across the full range of the planning process, from plan-making to development control and from central government policy guidance to local government initiatives. The EC tier of government is another policy-making body which British pressure groups attempt to influence. Furthermore, pressure groups are generally seen as a legitimate process of public participation though it is recognized that they may not be fully representative of public opinion.

9.3 THE MARKET

In Britain the industrial property market can be appropriately subdivided into three sections: the user market, the investment market and the development market. Within each section there is a further subdivision, as we identified earlier, for in some locations market behaviour determines property transactions while in others government seeks to improve economic conditions via an active industrial property policy. In the latter case private-sector activity is conditioned by public policy.

9.3.1 User demand

During the 1980s there have been fundamental changes in the location of industry, the numbers of employees and in the form of building required by industry. First, there has been a major industrial relocation in Britain away from older traditional centres of manufacturing such as the industrial cities of northern England. There has been a general relocation toward south-east England and away from other areas. There has also been a relocation towards small and medium-sized towns and away from the inner cores of cities; and finally there has been a general relocation towards sites which are easily accessible from the nation's major road network. As a consequence of these trends many small towns in south-east England which are near the motorway network have grown rapidly.

Second, the numbers of employees in manufacturing industry (the dominant user of industrial property) dropped during the 1980s from 6 800 000 to 5 100 000. This fall was most marked in larger manufacturing units with a consequent reduction in demand for large premises. This trend was partly offset by the growing quantity of capital per worker and the consequent reduction in the number of workers per unit of floorspace. Additionally, employers have become increasingly aware that skilled workers could be more easily attracted if the working environment was pleasant.

Third, during the 1980s, the design criteria which had previously been set by the investing institutions became increasingly influenced by users. Traditional developments continued though they have become technically more sophisticated (better insulation, facilities etc.). This perhaps reflects the harsher climate for industrial property developers and their wish to attract users via the development of more desirable properties.

Fourth, two new forms of industrial premises have appeared. High-quality units with a large proportion of office space (e.g. 60%) have been in demand. They have often been located in business parks or other environmentally attractive locations. In the later 1980s their development has been encouraged by the new Use Classes Order. The Stockley Park case study provides an example of this form of development.

Additionally there has been a growth in demand for very small industrial units (50–200 m^2). This has been associated with the growth of new businesses which has been promoted by central government. Supply to meet this demand has been provided by public agencies in depressed industrial areas and more generally by the conversion of older industrial units into multi-user premises by private owners.

All the above changes have been the result of movements in the demand for industrial property by users. Demand changes have led to rent changes which have altered the viability of development projects. The property developers are highly responsive to such changes, especially in the more prosperous parts of the country where rent levels have been sufficiently high to enable them to obtain finance for construction activity. The town planning system, with its increasingly market-led philosophy, has generally permitted such changes. In the more depressed regions private developers have been far less active though a number of good-quality schemes have been produced. In these areas, however, the public agencies have been active and generally have been quite successful at following market trends (if not leading them). This success must in part reflect the fact that the public agencies are dominated by the actors identified previously (particularly surveyors and planners), and market information within the professions has been transmitted quickly.

9.3.2 Access to the market

Users can enter the market as tenants, as owners, or as developers. For the aspiring tenant the key sources of information on availability, rents, service charges etc are chartered surveyors and public agencies. In areas of buoyant economic activity almost all of the industrial premises will be owned by private-sector landlords and the surveying profession will act as agents for them. The local authority will hold records of local chartered surveyors at either their estates or economic development sections. Users can expect rents per square metre to vary with building age, specification and location, and they should expect local economic supply and demand conditions to influence the final cost. Owners expect to have to negotiate with tenants over the final terms for the property and new entrants to the market should appoint their own surveyor to advise them on an appropriate negotiating stance in the current market.

In areas of economic decline the public sector normally plays a most influential role. The local authority or another public agency (such as English Estates) normally owns industrial premises which it wishes to let. The private sector still exists in these locations but the rents that it levies are heavily influenced by the property rents set in the subsidized public sector. In some locations the public sector is so dominant that private-sector development has not taken place for many years and consequently all of the private sector stock is second-hand. Again, access and information for tenants is best secured via public agencies and local chartered surveyors.

Becoming an industrial property user via purchase of an existing building is a common strategy. Purchasers may obtain the permanent (freehold) ownership of the property or a temporary 'ownership' of the buildings (e.g. for thirty years). In either case the purchaser will have to be willing to outbid any property investors who may wish to purchase the property and then let it to tenants. Moreover, an individual property on an estate of industrial premises it often not available. The owner would wish to hold all the properties so as to minimize management problems and also maximize potential redevelopment profits. That is not to say that premises to purchase are not available (they most certainly are), but it is not always possible to buy an appropriately sized premises in the right location and at an acceptable price.

Where the public sector is the major provider of new premises it is possible for the user to become the freehold owner of the property. Generally, public agencies have not been concerned with selling their properties to the investment market, though policy in this area does change from time to time. Users, generally, will have to pay a price which reflects the property's value in the market, and this price will be primarily determined by the investment sector. The public sector in these economically depressed areas produces buildings with public subsidy, often in advance of new tenants or owners. They provide the opportunity for a new entrant to the area to obtain premises quickly and cheaply, though it is not possible to guarantee the availability of buildings of the size and quality required by each user.

Generally, access to property via ownership would be best facilitated via local public agencies and the surveying profession. Legal advice on the full implications of property ownership should be sought from a solicitor.

To become a property user via the development of a new property is possible and may be the only feasible alternative for those seeking highly specialized premises. A useful, if slightly dated, source of information has been produced by the Department of the Environment (1982). Clearly the potential user-developer will need to consider the development process in essentially the same way as any other developer following the general procedure outlined elsewhere. Industrial property users should be aware that they will have to compete with other land users for sites. In recent years the change in the Use Classes Order has resulted in greater demand for land designated in local plans for industrial use. In areas of particularly buoyant demand for business premises, land prices have been bid up to reflect the demand for office-based commercial activity. Traditional industrial users have been outbid and they can find new sites difficult to obtain. The problem has become so widespread within parts of south-east England that some changes in the use classes may be made.

In areas where economic activity is more depressed there tends to be a surplus of land which has been designated for industrial property development. Local authorities, other public agencies and local surveyors will give advice on the availability of sites, and generally it will always be possible to find a suitable, serviced and attractive site. Several locations will exist where development receives some form of public subsidy (e.g. enterprise zones), though in practice this attraction may be partly offset by the higher price of land in these locations.

More generally, land prices will reflect potential development profits which themselves will be determined by the state of the national and regional economies. Land prices are volatile, and guidance on current local prices should be sought from an independent surveyor.

9.3.3 The investment market

Generally during the 1980s the institutions recognized that the major restructuring and relocation of British industry in the period would make industrial property investment risky, and they consequently withdrew from the sector. The private industrial property sector has become the preserve of property companies, who have been prepared to take the risk and undertake intensive management of the buildings, and owner occupiers. The only substantive exception to this trend has been the modern up-market business space developments where substantial expansion of demand has created real rental and capital growth.

The prices of industrial property investments reflect, as they do in the user market, the age and quality of the building, their location and the local conditions of user and investment demand. There are two additional factors: the current lease and the potential for redevelopment. The lease cannot be changed without the agreement of both parties. Clearly buying a property with a good tenant who has signed a 25-year lease with five-yearly rent reviews is a very different proposition from one where the lease has one year to run, the tenant is in difficult financial circumstances and where there are many empty adjacent properties making re-letting difficult. The impact of all these variables will be reflected in the yield (the relationship between rent and the price paid for the property), and advice on market prices and yields in what are often highly localized markets can be obtained from local surveyors. Equally these variables can be incorporated into a discounted cash flow calculation and an estimate of the appropriate price for the property can be obtained.

One variable which can substantially alter these calculations is the potential for redevelopment, for the value of an old industrial unit (its use value) may be quite unconnected with the current value of the site upon which it sits. Planning permission for redevelopment of a site to create new properties which are in the same use-classes division is usually little more than a formality (though planning permission is still required and major changes in design and density may be resisted). This opportunity may make the freehold purchase of the property very much more expensive than the use value would imply. What may be more important however is the probability of obtaining planning permission for a different form of development (e.g. demolishing an industrial building and creating an office development). The uncertainty created by Britain's discriminatory planning system means that the price paid by investors in these circumstances can best be regarded as speculative. Some forms of change of use are however more likely to be approved than others and detailed knowledge of planning policy (national and local) is necessary before estimates of value can be made.

The gains in land value associated with obtaining planning permission for redevelopment of land are subject to UK taxes but generally, and certainly at present, the tax regime is weak and considerable profits can be made from speculative property purchases. Equally considerable losses can be made if planning permission is not achieved.

For the investor it is possible to purchase single properties of various sizes. The town planning process in the post-war era has tended to encourage the zoning of industrial units into particular locations with the result that trading areas with 5 to 20 properties of various sizes are a more normal investment opportunity. In buoyant economic locations there will be an active market in such properties and bidding at an auction may be necessary. Advice on prices and the implications of investment in industrial property should be sought from chartered surveyors. In less buoyant locations there will be far fewer buyers and in the downswing of economic cycles it may be nearly impossible to sell property at anything other than very low prices. Yields in these markets are inevitably higher and reflect this issue. Regional policies such as tax subsidies and public sector property development activity can have sudden and unpredictable effects on the market. Overall, property investment in depressed areas can be described as highly risky but this certainly does not mean that it is not profitable.

While owner-occupation of industrial premises is common, the practice of being a tenant-occupier is generally accepted. The banking system does not require firms to own the premises they occupy and indeed a firm's bankers may well encourage it to sell their existing property and then let it back (known as a sale and leaseback arrangement), so as to create additional liquidity within the business. Property is seen, even by owner-occupiers, as an asset which can and should be exploited to an extent which would not occur in other countries. Even very large, multinational firms rent property in Britain. It is even possible for a firm to be taken over by another via the Stock Exchange so that the former can obtain the property assets of the later. This perhaps illustrates the extent to which property is perceived in Britain as an investment.

9.3.4 The development market

The process of property development in Britain can be complex, time-consuming and confusing. At its simplest the developer identifies a demand for a type of property, identifies a suitable piece of land, purchases the land, applies for and obtains planning permission, builds the property and then sells it or lets it to a tenant. In practice the process is complex.

First, identifying the demand for property is not easy. Pre-letting or pre-selling a development is not a legal requirement. The developer will attempt to estimate the rents and/or capital values that will be achieved when the development is complete, and in Britain's volatile property market this is not a scientific exercise. Valuation by comparison with similar properties is common. In a market where rent and capital values are rising the developer is particularly tempted to delay letting and sale in the expectation of a better return when the development is completed.

The second problem is to identify a potential site for industrial property development. This may be possible by studying local plans and discussing proposals with local planners and surveyors. If a site is identified within the local plan as suitable for such development then planning permission will probably be forthcoming. As stated earlier, it is possible to apply for planning permission for sites not included in local plans, and this approach is undertaken by developers because such land can be obtained more cheaply than the land identified in the plan. While it is not possible to generalize about the result of this process there is little doubt that, for some forms of property, it can result in considerable profits for the developer. In the industrial property sector, particularly for developments of the more traditional and simple design, this issue has not been of great importance, partly because the planning system has ensured that a large supply of suitable land is available (and therefore its price is relatively low), and partly because the planning system has not been willing to use its discretionary powers in the industrial property sector.

For new high-quality business space the issue is more relevant as the higher rents and lower yields make the whole process more profitable. Equally, occupiers prefer high-quality sites, often surrounded by open space. These are often unavailable according to local plans and hence some form of negotiation is seen as inevitable by developers.

The third problem involves purchasing the land. Buyers will have to compete in an open and competitive market and landowners are well aware of the value of land to potential developers. It is quite possible that landowners will attempt to exploit their ownership to obtain much of the profits of property development but in a location where there is a surplus of suitable land this is not usually a problem. It is possible that the local authority or other public agency owns the land, but these bodies do not usually purchase land by compulsory purchase as a means of facilitating private-sector development. They will also generally attempt to obtain the best price for their land. Where a public agency is the predominant supplier of new premises then land assembly by the agency is the norm.

Local authorities may be willing to enter into partnership agreements with developers, particularly where they are the owners of the site. Each agreement is different but the local authority will expect to have a long-term interest in the income from the property development.

The fourth problem associated with obtaining planning permission will depend on whether the land is included or excluded from such development in the local plan. Where the land has been identified as suitable for industrial development the local planning authority will generally accept the principle of such development but withhold full development permission until detailed plans are available. Planning authorities will be concerned with issues such as access, noise, smells, and waste disposal. It is not their intention to be obstructive in these circumstances but rather to ensure that any developments are built in accordance with local by-laws, are safe for owners and employees and do not create problems for immediate neighbours or the wider community. Where the land is not included in the local plan then additional issues will be considered by the planning authority. As stated earlier, planning authorities will need to be convinced that there are good reasons for deviating from their plan, for instance because there is now a shortage of industrial land or because the form and scale of proposed development was not foreseen when the plan was created. Planning authorities are not concerned with the potential profitability of the proposed development, nor are they concerned with protecting the value of existing developments by restrictive policies. They are there to decide whether the development would be in the best interests of the local community, and the local plan is the primary guide.

The fifth problem concerns the building process. Property developers in Britain may be construction firms but this is not necessary as the building activity can be contracted out to a construction firm. Advice on the various forms of contract for this process can be obtained from surveyors. The separation between development and construction permits very small development firms to exist in Britain; a firm with a total of less than 10 staff can handle a multimillion pound contract. This position is reinforced by the developer's practice of subcontracting out specialized development activities to other firms. The British tradition of professions facilitates this process.

The sixth problem concerns letting and/or selling the finished premises. The reasons for delaying the letting of a premises have already been considered. As the building nears completion the developer will be increasingly concerned to find tenants (and possibly long-term owners) and the advertising and promotion of the buildings will become of increased importance. The work is commonly passed to a firm of surveyors who have a key role in facilitating the market processes. At this time tenancy agreements have to be drawn up and negotiated with prospective occupiers.

It is possible that the building stage does not begin until a tenant or purchaser has been obtained. Clearly such a position will reduce much of the risk associated with property development. In the industrial property sector pre-letting is not uncommon for the building process can be completed very quickly and tenants will be prepared to wait, especially if the building can be modified to suit their precise requirements.

Overall the industrial property market in the more prosperous parts of the country is very much a free private sector with government only being involved through the regulatory powers of the planning system. In the more depressed areas the public sector is more heavily involved and often dominates the market (user investment and development) via its provision of subsidized premises.

9.4 CASE STUDIES

9.4.1 Stockley Park

This is a business park which has been developed since the mid–1980s and is perhaps the best British example of up-market high-quality industrial premises. The area was previously used for public waste disposal and was otherwise of a very poor environmental standard. It was also subject to fires created by the decomposition of household rubbish, and the local authority decided that development of a small part of the site could be used to finance reclamation of the whole site. From the

Figures 9.1 and 9.2 Commercial property in Stockley Park.

Figures 9.3–9.5 Commercial property in Stockley Park.

developer's point of view the site was almost ideal, being close to Heathrow Airport and major motorways. It had obvious attractions to international companies, and if a high-quality site could be created it would be attractive to managerial and highly skilled employees.

Development of the site was restricted by the fact that it is in the Green Belt, and consequently received the highest level of protection available from the planning system for an undeveloped site without historic or scientific significance. The developer managed to persuade the local planning authority that it would be in the best interests of the local community if a substantial building programme was permitted even though it involved a major departure from conventional planning policy. Planning gain was to be provided in the form of an 18-hole golf course, sports fields, a dramatic improvement in the environmental quality of the whole site and public access to large parts of it via footpaths and open spaces.

Work on the 140 ha site started in 1985 with a major process of earth-moving and site reclamation. A total of 40 ha has been allocated to the business park, which is still being developed, and the remaining land has been improved to provide a high-quality environment. The developer has provided shell and core buildings leaving the occupiers to fit out the premises to their individual specification. The buildings are provided on a leasehold basis, are of two or three stories, have generous ceiling heights for air-conditioning systems, and are provided with ample car parking. One particular feature of the site is the extensive use of water as a means of creating an attractive environment.

Over 60% of Stockley Park is occupied by non-British firms; it accommodates company head offices, research and development centres and computerized production facilities. The emphasis is on skilled staff, technology and decision-making. To ensure that the site is maintained at the highest possible quality the developer provides an active site-management team. In addition to maintaining security and environmental quality the management team oversee the provision of sports, retail and banking facilities for the employees located in the park.

Stockley Park illustrates two clear points. First, if a development proposal is seen to be in the best interests of the community, then it is possible that planning permission can be obtained even where the development implies the overturning of one of the most 'sacred' planning policies. Such a decision may be made at local level but central government will still have to give its approval. Generally these exceptions to the local plan and/or national planning principles (of which the protection of the Green Belt is but one) result in planning inquiries, because either the local authority refuses development permission or pressure groups effectively lobby the Secretary of State for the Environment to call such an inquiry. He may overrule any objection after receiving the results of the planning inquiry and in this case it is assumed that the government perceives a nationally important reason for approving the development application. It should also be noted that the Minister can even overrule the advice of the planning inspector. Clearly this process injects some uncertainty into the planning system but it must be remembered that these overriding decisions are the exception rather than the rule.

Second, Stockley Park illustrates (in an extreme form) the type of development that has become increasingly common during the 1980s in Britain. While traditional industrial units are still being constructed, especially in older industrial areas, the growth sector of the market has been the better-quality premises with a high proportion of office space. This change has been facilitated by the creation of the new Use Classes Order but, more importantly, it reflects the changing nature of industrial property demand.

The planning system in the 1980s has been market-led, and where there is a perceived market need for a new type of property the planning system has been able to respond to the need. Also, one of the benefits of Britain's discretionary planning system is that it can respond rapidly to new demands because there is no immediate need to change the formal plan.

9.4.2 Team Valley

The Team Valley Trading Estate has its origins in the 1930s when the government decided to respond to the high levels of local unemployment by an interventionist policy of industrial property development. The first site chosen was in Team Valley, on the south side of the river Tyne and approximately 5 km from the centre of the Tyneside conurbation.

The site of 280 ha was purchased from local landowners and passed to the government agency English Estates. It was seen as particularly suitable because of its size and its accessibility by rail and river transport. In later years the development of the road network also favoured the location. The site included some marshland, and a general programme of site improvement was needed before building could begin.

The intention was to provide factories in advance of demand, to a high specification and in a pleasant environment. It had been noted that such estates were being developed in the more prosperous regions of the country by the private sector but in the depressed north of England no such private initiative was likely. The industrial property was not only built in advance of demand but with the intention of attracting mobile firms from other parts of the country.

The site has been the focus of development and redevelopment over its life, and the traditional form of simple industrial unit has been the normal form of development. Subsidies have varied over the period and indeed during the 1980s. One common practice of English Estates has been to build and let properties at rents which reflect the 'current market rent' in the locality. Given the high yields for industrial property in the area and the generally low level of rents, this has implied building property which has a market value somewhat less than the costs of constructing it. The private sector, naturally, has been unable to compete with this subsidy and generally public-sector development on this and other sites dominates the region. English Estates not only lets property but also sells it to owner-occupiers and even sells serviced land to private developers where there is some strong private demand.

English Estates' development programme also influences the property-investment market in the region, for investors perceive that continued development is likely to reduce the possibilities for rental growth.

This is one reason why the yields for industrial property in the region are so high.

During the 1980s part of Team Valley has been designated as an enterprise zone. Developers are deemed to have planning permission for developments which are to be built in accordance with the rules for the zone and development receives tax subsidies during the period 1981–1991. Development within the zone has been rapid during the period. A substantial retail park has been developed and much of the remainder of the zone has seen the creation of a high-quality business park. English Estates has serviced the land in the site and generally sold it to developers. There is evidence that some local firms have moved their business a few miles to take advantage of the zone but equally there has been an influx of new business to the region.

During the 1980s English Estates have been particularly active in encouraging the creation and expansion of new businesses. They are able to give business advice to their tenants and have also provided a number of small business units. On some occasions these small units have been located together and business services have been available on a shared basis. One often-overlooked policy of English Estates has been their willingness to offer short-term leases rather than the normal 25-year term. This has made it easier for new businesses to obtain premises, for they do not have to commit themselves beyond the current outlook for their business.

The public agencies which provide industrial land and property and particularly English Estates have been recognized as innovators and initiators in industrial property development. They have been particularly active in attempting to find new ways of developing industrial sites and new forms of property. For a public agency this is quite surprising, but they have been staffed by professionals with a wide experience of the property market and they act as an independent agency rather than as a government department.

Team Valley can be seen as an example of publicly subsidized property development. In an economically depressed region the provision of industrial units in advance of demand has been a major element in the regional strategy for regeneration. Clearly the existence of such public activity will have significant implications for private property investment and development.

9.5 CONCLUSION

The industrial property sector in Britain can be viewed as a highly competitive market, responsive to changes in user demands via price, quality and locational adjustments. The sector is naturally affected by public policies and equally by market pressures on land use by other sectors.

The planning system, while in principle discretionary, has generally in the industrial property sector ensured a sufficient supply of land for development and ensured that the actual process of property development has taken place within plan guidelines. During the 1980s the planning system has become more market-led with the consequence that even greater effort has been made to ensure that plans are in accordance with the needs of industry. However, in areas of buoyant demand, there are inevitable conflicts in land-use allocation, and an adequate supply of land cannot be guaranteed. In these locations there is room for speculative applications for development approval on sites which are not included in the local plan.

In areas of economic depression, which for industrial property includes a large part of the British Isles, the key public issue is not land availability but rather the fact that commercial development of industrial space has not been viable and public agencies have been actively involved in the creation, letting and sale of industrial property on a scale which has led to public-sector domination of the market. The goal of public policy has been the moderation of regional disparities in unemployment though in the 1980s the policy stance has been directed more towards improving efficiency and competitiveness. Private markets do exist in these locations but they are primarily second-hand markets, and rents and values reflect the levels created by the subsidized public sector. The public sector agencies, particularly English Estates, are very market-led and make determined attempts to produce property in accordance with the demands of users.

Access to industrial property can be via ownership or tenancy. In either case the process is not complicated but potential users should ensure that they are aware of all the obligations and rights associated with each form of tenure. Legal advice is essential.

The property markets in Britain tend to be cyclical and access problems can occur at both the peak and trough of the economic cycle. In times of depression, finance for property purchase may be difficult to obtain, and new premises will be in short supply. In boom times finance will be easier but the high level of demand may outstrip development activity. Good planning and preparation before attempting to secure property are advisable.

The various sectors of industrial property, the user market, the investment market, the development process and the land market must be seen as one highly interconnected market where property values substantially determine land values and development activity. With fully flexible and uncontrolled rents and values there is every incentive by owners and occupiers to search for the best alternative within the market. For occupiers this can involve seeking buildings with appropriate facilities at acceptable rents and in suitable locations. For investors it implies that prices paid for premises should reflect the market's expectations with respect to rent and capital value growth over the life of the property. For developers it implies seeking out viable development opportunities where property can be created at a cost which is less than its value as determined by user and investor demand.

Demand for industrial property in Britain has evolved in the 1980s. While many traditional industrial buildings are still created, the most profitable form of development has been the up-market business unit where a high quality specification and a large proportion of office space predominates. Inevitably in a market-led system this has resulted in land designated for industrial property development being used for office, research and distribution/communications functions rather than manufacturing activity, and it has resulted in a shortage of sites for new traditional property types in certain locations.

The locational characteristics of demand have also changed in the 1980s with a movement towards the south-east of England, away from the centres of major conurbations and towards smaller, free-standing towns with good access to the country's road transport system.

The British system of regulation of land use, and the market-led process of property development, investment and usage appear to be reasonably consistent with one another. Problems do occur where demand for developable land is intense, but this is perhaps inevitable. The British solution of effectively permitting industrial land usage to be increasingly determined by market pressures has left a legacy of imbalanced provision of premises, and has inadequately co-ordinated infrastructure provision.

Away from the areas of greater market pressures, an equally vital issue is how the public sector should respond to the need for industrial property development in regions where the level of demand makes commercial property development non-viable. The British response of creating an active public agency with considerable public subsidies has stimulated substantial activity but it has also had the effect of eliminating the local development market and has made it very difficult for even a radical right-of-centre government to return the property market to a market-based system.

REFERENCES AND FURTHER READING

Board of Inland Revenue (1989) *Capital Allowances on Industrial Buildings and Hotels.*

Brindley, T., Rydin, Y. and Stoker, G. (1989) *Remaking Planning: The politics of urban change in the Thatcher years*, Unwin Hyman, London.

Cadman, D. and Austen-Crowe, L. (1990) *Property Development*, E. & F. N. Spon, London.

Davies, H. W. E. (ed) (1989) *Planning in Western Europe*, HMSO, London.

Department of the Environment / Department of Industry (1982) *The UK Construction Industry: A Guide to Methods of Obtaining New Industrial Buildings in the UK.*

Department of the Environment (1984) Circular 16/84, Industrial Development.

Department of the Environment (1987) The Town and Country Planning (Use Classes) Order.

Department of the Environment (1988) The Town and Country Planning General Development Order.

Department of the Environment (1988) PPG4, Industrial and Commercial Development and Small Firms.

Fothergill, S., Monk, S. and Perry, M. (1987) *Property and Industrial Development*, Hutchinson, London.

Goddard, J. B., and Champion, A. G. (eds) (1983) *The Urban and Regional Transformation of Britain*, Methuen, London.

Jones Lang Wootton (1991) *A Guide to Office and Industrial Rental Trends in England and Wales* The 50 Centre Guide, London.

Monk, C. (1988) *Science Parks and the Growth of High Technology Firms*, Croom Helm (in association with Peat Marwick McLintock), London.

Morley, S., Marsh, C., McLintock, A. and Martinos, H. (1989) *Industrial and Business Space Development*, E. & F. N. Spon, London.

Waldy, E. B. (1986) *Business Parks* Fletcher King, London.

STATISTICAL APPENDIX

Statistical information on the industrial property market is not generally collected by either central or local government. The last major government study of commercial and industrial floorspace relates to the period 1982–1985, though some local authorities do have more up-to-date information on their locality. The Valuation Office, which is part of the Inland Revenue (the tax-collecting authority), publishes a *Property Market Report* annually but this provides only a limited amount of information at an aggregate level.

The main sources of information are the local property surveyors who collect information on industrial property as part of their service to their clients. Local information is essential for both occupiers and investors for the levels of rents and property values are determined, as the text makes clear, by the local conditions of demand and supply. The location, accessibility and the environment in which buildings are situated all affect the property's value. Within a distance of 5 km rents and values can change by more than 50%.

It is perhaps inevitable that those who have access to the information perceive it as an asset and they are usually not willing to divulge current market tends. Information becomes less valuable over time as it becomes more available in the marketplace and then it may be published by the surveyors in the local press and in their own 'in-house' publications.

The importance of *local* information explains why small firms of surveyors can compete with national firms, for a national firm will only be able to keep abreast of local conditions if it has an active local branch office. National firms are, however, able to produce comparative data from their branches and much of this is published, after the lapse of a suitable period of time.

Typical data are provided in Tables 9.1 and 9.2. The regional data are useful for researchers but it is the town data and even more disaggregated data which are most useful for occupiers and investors. Generally the data that are collected reflect the needs of the market, and relatively little research is directed towards issues which are of interest to the wider community.

Table 9.1 Regional statistics

Region	Rental growth (%)	Rental growth[2] (%)	Yield[3]	Available floorspace[4]
London	19.2	13.0	7.9	1.42
South East		7.9	1.62	
South West	20.8	12.8	8.5	0.30
East Anglia	20.4	13.5	8.4	0.17
West Midlands	12.6	10.7	8.5	0.84
East Midlands	21.5	12.4	8.5	0.75
North West	9.3	8.7	9.3	1.04
North	21.8	10.4	10.1	0.40
Yorkshire and Humberside	14.6	11.3	9.3	0.96
Wales	5.5	9.8	9.1	
Scotland	20.0	7.4	11.0	

[1] Rental growth for the one year up to June 1990.
[2] Annual rental growth, (compound), since 1977.
[3] Yields as of end of 1989. Figures express full rent divided by capital value.
[4] Floorspace available in millions of m², as of April 1990.
(Sources: The national surveying firms of Healey and Baker, Hillier Parker, and King and Co.)

Table 9.2 Town statistics rent levels

Town	Region	Business premises[1] (£/m²)	Industrial premises (£/m²)
Chiswick	London	193	86–91
Hounslow	London	183	97–102
Basingstoke	South East	129	70–75
Portsmouth	South East	108	76–78
Bristol	South West	118	62–70
Swindon	South West	108	57–65
Ipswich	East Anglia	97	54–59
Coventry	West Midlands	75	40–46
Derby	East Midlands	62	40–46
Manchester	North West	86	32–40
Newcastle	North	65	35–40
Sheffield	Yorkshire and Humberside	91	46.48
Cardiff	Wales	65	44–48
Glasgow	Scotland	102	38–44

[1] Business premises may include up to 50% office space and are generally of a higher level of specification than industrial premises.
[2] Industrial premises may include up to 10% office space.

10

SOME COMPARISONS AND CONTRASTS

Barry Wood and Richard Williams

In many respects this is the type of book which invites the reader to extract from it whatever concepts and information are useful to them in respect of the country or countries with which they are most concerned, and to draw their own conclusions. It does not purport to be a comprehensive comparative analysis. Therefore, this concluding chapter is not intended to offer a systematic review of all the issues raised in all seven countries covered. Rather, it aims to draw attention to a selection of issues and examples from the different countries in order to suggest some comparisons, similarities and contrasts or significant differences.

Recognition of the differences, especially where they are conceptually founded, is vital not only to an academic analyst of the subject but also to the practitioner. It is easy to recognize specific differences in procedure and regulation when one has in mind the model of one's own system of planning and property law, whose complexities are well recognized and understood. From this model one looks for, and usually finds, equivalent complexities in the other countries systems which need to be given close attention. This constitutes a relatively straightforward, indeed simplistic, view of the problem of transfer of technical and professional skills from one country to another. It may work when the basic concepts and thought patterns underlying the law, procedure, and professional practice of the country being studied are the same as those of one's own country.

However, the countries of western Europe described here present a variety of legal, technical and professional practices and procedures, reflecting differences of constitutional arrangement and culture that are deeply rooted in the countries concerned. It is often necessary to appreciate these differences in order to make the switch from operating in one country to another successfully as a practitioner. It is essential for the prospective professional practitioner in another country to recognize that there are often complex features of the system in one country which may not correspond with any complex issue in another; that an important issue in one country, to which a great deal of professional attention has to be given, may be a non-issue in another. The person coming from the country where such an aspect of the subject is a non-issue would be in danger of making mistakes or misjudgements if they were not on the look-out for complexities which do not arise in their own country.

Likewise, misunderstanding may arise because particular functions or services operate in the public sector in one country and private sector in another, or within quite different types of organization. Responsibility for comparable valuation data, held by public-sector valuation exchanges in Germany, is a case in point.

10.1 OBJECTIVES OF CROSS-NATIONAL STUDY

Much has been written, especially by planning academics, about the objectives of cross-national comparative study, and what realistically can be expected of such work. Two basic objectives are well established: the improvement of planning practice and the advancement of planning theory (Faludi and Hamnett, 1975). Faludi and Hamnett also proposed a third objective: that of working towards the unification of the field of planning. At that date this may have seemed an abstract and theoretical preoccupation; nowadays with the imminence of the completion of the Single European Market, it raises questions for those operating in a practical sense in the industrial or any other property market multinationally within the single market, to which this chapter returns later.

In the context of industrial property, the improvement of practice is a clear reason for studying other countries of interest. This can be in order to learn to operate in other countries, or in order to learn from the experience, institutions and procedures of another country with a view to improving practice in one's own country. The experience of those involved in cross-national comparative studies in planning has often been that the benefits in the form of providing the basis for a critical appraisal of one's own system from a new perspective are considerable, but that the idea that one can neatly transfer procedures, policies and practices from one country to another often proves to be naïve, as the complexities involved are often underestimated.

Planning and real estate are not subjects in which a kit of parts can be offered enabling practitioners to convert from one country to another. The fundamental lesson to be learned is that it is over-simplistic to look for equivalents in another country of specific procedures and complexities familiar in one's own country. Instead it is necessary to comprehend the logic, the thought patterns and underlying concepts in the other country in order to appreciate of how to operate professionally in that country. Hence the significance of the second basic objective of comparative study, namely to improve theoretical understanding. It is not unusual for practitioners in planning and property development to dismiss the practical benefits of a theoretical understanding of the subject, but experience suggests that it is very necessary for these reasons.

The study of other countries often tends to lead on to theoretical development because the need to understand different concepts and their underlying logic becomes apparent: the wider constitutional and cultural contexts must be explicitly noted rather than taken for granted as easily happens in one's own country, and it may be possible to evaluate newly proposed policies on the basis of comparable experience elsewhere. For example, if a government proposes some new concept on the basis of claimed theoretical advantages (as was the case with enterprize zones in the UK) it may be possible to test the theory by comparative study, thus advancing theory to some practical purpose. The theoretical and conceptual objectives of comparative study may therefore provide an effective foundation not only to build an understanding of specific details to which attention is drawn, but also to predict how the system might operate in situations not encountered before, and to recognize complexities which have no equivalent in one's home system.

Often, the question is posed by material such as that contained in the preceeding chapters: 'Why is there no such-and-such in country X?'; 'How can you manage to operate professionally in your real estate markets without Y?' Both theoretical and practical understanding are necessary to answer these questions satisfactorily. For example, in discussion about European real estate systems, the question has several times been posed to the editors: 'How can you manage in the UK without a cadastral survey?' Equally, one could ask of the Netherlands, 'Why is there such a small investment market in industrial and commercial property?'

The trick, when learning to operate professionally in another country's property market, is to turn these questions round and pose the question 'What are the specific features and complexities that I need to understand in order to operate in this market, which have no equivalent in my home market, and which therefore cannot be foreseen by assuming equivalence with the basic model of my own system?' To answer such practical questions for oneself, a conceptual appreciation of the underlying logic of the other country's planning system and real estate market must be acquired, which in turn requires some understanding of the constitutional and cultural context, and of course of the language.

The third objective of cross-national study referred to above was that of unification of the field of planning. Williams (1984; 1986a) has suggested an alternative formulation orientated more towards policy-making: in terms of seeking a common understanding of the nature and variety of planning systems at the level of detail necessary for the formulation of policies at the supra-national scale. This could apply to supra-national jurisdictions such as the European Commission, or to commercial firms formulating their strategy for international operations.

In the context of the SEM and European integration, the question of unification or harmonization of a field of activity, or of learning to handle, understand and operate within the variety and complexity of national practice that is found in the European Community, has to be faced by those operating in planning and industrial property and those responsible for formulating any future EC legislation regulating these operations, relating directly back to the third objective of cross-national study briefly discussed here. In practice, the second option, that of learning to handle the complexity and variety, is the one that is in effect being pursued to date.

A fuller discussion of the many conceptual, methodological and practical issues raised by cross-national study and research can be found in Masser and Williams (1986). Issues of harmonization, convergence and learning to handle complexity will be returned to later in this chapter, following a review of topics in the preceding chapters. This review is intended to be illustrative rather than comprehensive.

10.2 FLEXIBILITY OF PLANNING SYSTEMS

The distinct difference between the British (and Irish, although outside the scope of this book) discretionary system of planning and those of other countries comes through clearly in the national chapters. A fundamental means of achieving flexibility in the UK planning system is the planning authorities' discretion to decide whether to grant planning permission on the merits of a proposal, without this being predetermined by legally binding plans.

The requirement for an approved plan is an essential prerequisite for the authorization of development in the other countries, where plans may be legally binding and there is a direct and formal relationship between the plan and authorization to develop: at its simplest, the model elsewhere is that development specified in the approved plan must be authorized, and no other development could receive authorization. On the surface, this would appear to generate great inflexibility both for developers and for the planning authorities, and more generally would tend to limit the ability of the development planning system to respond adequately to changes in market demands. In practice, however, it is always far from being as simple and rigid as that, as each country has devised many methods, either within planning regulations or otherwise, to achieve flexibility and responsiveness to changing market conditions.

The French, for example, by the designation of a ZAC are able to respond quickly to development opportunities, while the Spanish achieve flexibility by limiting the degree of precision with which land uses are defined, and by use of the facility for designation of activities as being of 'social interest'. In principle, the legally binding nature of the Dutch and German building plans (*bestemmingsplan* and *Bebauungs-plan*) offers very strict control. Indeed, in Germany statutory plans are not merely approved in accordance with the law: they *are* the law.

Nevertheless, this apparent rigidity can be overcome, and flexibility is achieved in a variety of ways without impeding unanticipated but acceptable development. A legally binding system, founded in Roman law concepts, may appear rigid, but flexibility and market responsiveness are achieved by specific means taking advantage of detailed provisions in legislation.

The UK's discretionary system facilitates, but is not a necessary pre-condition for, the market-led approach to property development which has been characteristic of the UK's property market during the 1980s. The UK government has in 1991, following the passage of the Planning and Compensation Act, indicated a return to a more plan-led approach to development policy for the 1990s. This will bring the UK closer to the position in Germany or the Netherlands. Equally, however, a system of legally binding plans does not necessarily indicate a plan-led system, since in Italy for example extensive negotiations occur in a market environment, with the formal plan being prepared after development proposals are agreed in order to grant legal authorization to the develop-ment. In practice, in all countries, the significance attached to the policy-making process, the degree to which its outcome is adhered to, the extent of informal procedures and the freedom with which market forces can operate are matters of political will and the political environment surrounding the planning and development process.

The increasing cross-national competitive pressures released by the creation of the Single European Market are likely to encourage the development of a more flexible approach as each country seeks to attract mobile capital. This may in due course lead to some *de facto* convergence of the systems.

10.3 LEGAL SYSTEMS

The distinction which has been referred to between the UK's common-law system and the Roman law system found elsewhere, and which broadly correlates with the distinction between legally binding and discretionary planning systems, is important but must not be overemphasized. In fact, within the scope of this book three rather than two legal traditions can be identified. One is the strictly codified form of Roman law to be found in France (with the *Code Civil*) and Germany (with the *Bürgerliches Gesetzbuch*, BGB, 1896 – civil code). The second is the common-law system of England and Wales (Scotland, though united with England to form the UK since 1707, still has its own legal system). The third legal tradition (exemplified by Sweden) occupies a half-way house between these two.

The Swedish system is based on old Germanic legal traditions, with influences from Roman law and features drawn more recently from French, German and Anglo-American law. Sweden abstained from large-scale codification in the French or German manner but it is largely written law, with case law playing a small but important role. In Scotland, the system is essentially Roman in origin, to which legislation developed under the English common-law system has been adapted. Sweden and Scotland therefore both represent something distinct from the classic Roman and common-law systems, although elements of both are to be found. It is suggested therefore that they can be placed in this third category.

10.4 THE SPEED OF THE PLAN-MAKING PROCESS

Several authors have pointed out the slowness with which the plan-making process functions in their countries. It can seem interminable, particularly from the point of view of the property developer. However, there is no evidence that a slow pace of formal plan-making and plan approval necessarily implies a slow pace of development, or even necessarily a slow pace of authorization of development. Furthermore, ways and means exist to expedite these procedures in all countries whenever the political will exists to do so.

A true measure of the speed of the process requires more detailed research than is available now, in order to ensure accurate measurement and comparison of like with like. Such a measure would need to define the real start and end point of the process, to include negotiation prior to the formal plan-making or development control phase, formal authorization and implementation of the development. It is suggested on the basis of experience, rather than such precise research, that the true length of the planning process in different countries does not vary all that greatly.

In several cases a form of policy-planning occurs either within the framework of the planning system at a different level of government or spatial scale (by the Dutch provinces and national government planning service, the *Rijksplanologischedienst* for example), or as a separate and possibly less formal process of policy-making for economic development.

Additionally, of course, vital discussions and decisions concerning development proposals within many urban areas occur informally within political or other groups and meetings. Thus, by directing attention at the formal town-planning procedures, including the plan-making process, one can miss important policy-making processes. Formal plan-making procedures can be long and drawn out, or they may be speedy rubber-stamping processes following a long period of behind-the-scenes discussion of a project. Either way, all can move quickly when it is politically desirable to do so.

The levels of the hierarchy of government at which planning policy is formulated and development decisions are made also constitute an important variable, as does the possibility of legislation affecting the industrial property market being enacted by sub-national levels of government. In the UK, the position is simple: there is no legislative power below that of central government, and most planning decisions are taken at the municipal level of government. It is not only the wholly federal systems such as that of Germany where power to legislate exists at a sub-national level. Such powers may also be exercised under regional devolution arrangements in Italy or Spain, for example. Policy frameworks and strategic development plans may also be developed at different levels. In the Netherlands, for example, this can occur at national and provincial levels, while in Germany it can occur at state and sub-regional levels.

Where regional government or representative authorities exist, there is usually some form of regional policy, normally incorporating policy for economic and industrial development. Incentives for regional industrial development are frequently associated with this regional level of government, rather than existing as part of national policy as in the UK. For a fuller discussion and listing of regional incentives for industrial development, see Yuill and Allen (1990).

10.5 LAND SUPPLY

Within each country there is a process by which land is supplied for development. Practice is diverse, however. The Dutch system involves an explicitly public-sector land-assembly process which ensures a supply of land at controlled prices. Most other countries make at least some use of private land-assembly and, where this is inadequate, resort to public intervention. This has impacts on land prices but the form of impact appears to be influenced by the form of intervention. In Britain the history of the UDCs is informative because it shows how public land assembly, coupled with other public policies, can result in very high land prices. By comparison the Dutch form of intervention appears to ensure that land is both cheap and available in the right locations.

10.6 TENURE

It is apparent that there are wide differences in the forms of tenure between the countries considered in this book. Each country defines the rights associated with owner-occupation and the rented sector with some precision, and the practitioner clearly needs to grasp both the details and their significance for market behaviour.

The difference between the industrial property market in Britain, with its important role for property investors, and the norm of owner-occupation in other European countries, is well known. What is more revealing is the extent to which rented tenure is becoming more important in other countries. Is this a result of the growing influence of British property professionals? Is it being demanded by occupiers who have experience of the tenure elsewhere and now see it has advantages for them? Or is it the result of some fundamental shift in the structuring of property ownership within the EC? The answer is not clear, but the issue of tenure is the one area where we see real evidence of harmonization in the property world. It is more than conceivable that within a decade renting or buying an industrial building in Newcastle, Nijmegen or Turin will involve essentially the same obligations, responsibilities and benefits. The differences could be little more than cosmetic.

The growth of rented tenure does of course imply the growth of a property investment market. In the UK such a market is well established and those involved are used to the way it functions. In other countries property investors may not behave in the same manner as those in Britain, for they may be driven by different objectives and time horizons. In practice, however, we expect market behaviour to differ little between countries. National regulations may have some impact, but so many of the investors are international organizations that we are driven to the view that they will buy and sell property in essentially the same manner in each market. As time and experience develop, Europe is likely to become one integrated property investment market and, at least for the best properties, the price will be evaluated by the same procedures.

10.7 PUBLIC SUPPLY OF INDUSTRIAL BUILDINGS

It is a characteristic of most countries that there are public interventionist policies to ensure a satisfactory supply of developable land on a financially viable basis to overcome inter-regional disparities in areas or regions suffering from economic decline, peripherality or underdevelopment. In the Netherlands, the system of land supply goes further, by ensuring a regular supply of development land at consistent prices. The UK is the exception because public policy in the 1980s has consistently sought to minimize public expenditure; there is no effective regional lobby or influential political voice for the regions, and therefore uniquely the regions of the UK are treated as outlying fiefdoms by central government. In every other case reviewed, some form of regional or provincial government or representative authority exists, and a public role in regional policy is to be expected, providing an important context for industrial development.

One issue addressed everywhere is the need of at least one tier of government to become involved in the provision of industrial property within regions that suffer from economic depression. Anyone familiar with the situation within Britain would be able to grasp the policy framework within Sweden because the issues identified and the solutions proposed have remarkable similarities. Governments have been unwilling to leave the issue to the markets because they are well aware that little property development would take place and that there would be little hope of eliminating regional disparities without intervention.

Advance factories or industrial sheds built, normally by a public sector agency, in advance of the identification of any occupier as a regional economic development measure have been a feature of UK public policy for many years. The leading roles have been taken by English Estates and the Scottish and Welsh Development Agencies. This concept has rarely been adopted elsewhere, and for this reason was not readily accepted at one time by the European Commission as an acceptable application of ERDF funds. It is interesting to note the interest of certain German municipalities such as Cologne in this concept, although such factory building does not figure in Federal economic development programmes.

10.8 EC REGULATION

It is notable that most contributors have identified the EC Directive on Environmental Assessment of Projects (Directive EEC/85/337) as being significant, and have described how it is incorporated into the planning and land-use authorization process. National regulations implementing this Directive took some time to be adopted, and do not all adhere to the same environmental thresholds and standards. Furthermore, there are still questions over the extent to which compliance is achieved in all countries.

Nevertheless, this Directive represents the first step in the process of elimination of pollution havens and the creation of the level surface throughout the EC on which locational decisions by a hypothetical mobile multinational industrial concern would be taken (see discussion in Williams, 1986b). It is therefore an essential component, though only a first step, in the completion of the internal market and the creation after 1992 of an integrated market for industrial property development.

Chapter 2 raised the question of harmonizing authorization procedures and the other rules and regulations governing real estate and property development, following the logic of the SEM and the removal of non-tariff barriers as part of the 1992 programme to complete the internal market. The evidence of this book is that the variety of existing procedures, when looked at in detail, is immensely complex and that any harmonization procedure would be difficult and time-consuming. The alternative scenario, which is being followed so far, involves the development of a core expertise able to handle the complexity of the different systems rather than harmonization into one system. However, this may in the longer term prove to be merely an interim solution – a stage in the road to European integration.

Much has been made of the different regulations and procedures in the different countries, but the multiplicity of professional titles under which the many actors in the industrial property markets operate, with their widely varying status and esteem, also poses acute problems. Questions of what all the titles signify, what level of education, qualification and experience can be assumed, and of the extent of mutual recognition all need to be tackled. The biggest anomaly in the property field and real estate professions is the lack of correspondence between the UK's Royal Institute of Chartered Surveyors and professional associations in all EC countries outside the UK and Ireland. The RICS is taking initiatives to promote a wider understanding and recognition of its qualifications, expertise and professional ethics within the EC, but it is far from certain that this will find ready acceptance in member-states with different traditions and professional structures.

10.9 CONCLUSION: LEARNING FROM ONE ANOTHER

Each country portrayed here has a planning and industrial property market which, when looked at in detail, appears quite unique. At one level it is hard to believe that anyone has learned anything from one another, or that policies that are useful in one country could be of benefit to others. At another level, there are clearly common features and evidence of convergence of ideas and practice. It may be helpful here to consider planning and property development separately.

In planning, the issue appears most starkly. Planning systems are national-specific, bounded by national laws and customs and directed towards a policy agenda largely defined from a national perspective. However, planning design concepts in the context of industrial development, science parks, etc, have become international. Clearly many problems faced by public policy-makers are common throughout Europe and it is here that there is now a real opportunity to learn from the experience of others.

For example, the UK could look more closely at the benefits of legally binding plans, especially as a more plan-led approach is now declared to be government policy. Equally, questions of discretion, certainty and how flexibility is achieved should be posed to other countries from a UK point of view. It is not appropriate that we should all move to some idealized system as national requirements will dominate the planning system for the foreseeable future. Harmonization of planning systems and procedures is not advocated, as it would be incredibly complicated to achieve, but some convergence of practice is likely to be worthwhile. In terms of planning policy there is collectively in Europe a wealth of experience that could be shared for mutual benefit.

In the property markets themselves there is more evidence of learning from one another, as the leading property firms and property professionals increasingly operate internationally. Tenure, investment, even the built form – all show signs of influences from outside. In the rush to attract mobile capital and to maintain national attractiveness to existing resident industry, property-market professionals are ensuring that occupiers are faced with similar opportunities everywhere, and governments are co-operating with the process of change that is required. More learning from each other could undoubtedly occur: for example, Italy and Spain could surely learn something from industrial property development in Britain by English Estates.

Nevertheless, in the particular sector of the property market reviewed here there is limited evidence of deliberate learning from each other, in the sense that one country has sought to import procedures and practices from another, although there is substantial scope for this to occur. A process of learning from one another is therefore happening through 'push' factors, or the export of experience, rather than 'pull' factors or the deliberate importing of experience.

The simple transference of ideas and concepts from one country to another is fraught with risks for those who do not understand the frameworks and legal and cultural environments within which the planning systems and industrial property markets function.

A comparative study usually leads to a heightened appreciation and more critical appraisal of one's own system from a newly acquired perspective, and to an ability to stand apart from it. We are sure that this book fulfils this role. This will certainly be a valuable skill as we are witnessing at present in Europe a situation where different national systems and concepts are competing in each other's territory for custom and dominance. No clear picture yet emerges of who or which will prevail, or whether some form of coexistence or convergence will be the outcome in post–1992 Europe.

REFERENCES AND FURTHER READING

Faludi, A. and Hamnett, S. (1975) The study of comparative planning. Paper CP–13, Centre for Environmental Studies, London.

Masser, I. and Williams, R. H. (eds) (1986) *Learning from Other Countries: The cross-national dimension of urban policy-making*, Geobooks, Norwich.

Williams, R. H. (1984) Cross-national research: translating theory into practice. *Environment and Planning B*, 70 (1), 149–161.

Williams, R. H. (1986a) Translating theory into practice in Masser and Williams (1986) (*op. cit.*), pp. 23–39.

Williams, R. H. (1986b) The EC environment policy, land use planning and pollution control. *Policy and Politics*, **14** (1), 93–106.

Yuill, D. and Allen, K. (1990) *European Regional Incentives: Directory and review of regional grants and other aid available for industry and business expansion and relocation in the member-states of the European Community and Sweden*, Bowker Saur, London, in association with the European Policy Research Centre, University of Strathclyde.

INDEX

For Product Safety Concerns and Information please contact our EU representative GPSR@taylorandfrancis.com
Taylor & Francis Verlag GmbH, Kaufingerstraße 24, 80331 München, Germany

www.ingramcontent.com/pod-product-compliance
Ingram Content Group UK Ltd.
Pitfield, Milton Keynes, MK11 3LW, UK
UKHW022014180926
475701UK00003B/360
* 9 7 8 0 4 1 5 5 1 1 8 9 6 *